WILEY

本书出版受以下资金或项目资助：
国家自然科学基金（编号：31860233）
江西师范大学教改课题（编号：JXSDJG1812）
城市建设学院教学经费

Ecology of
Urban Environments

城市环境生态学

[澳] 柯尔斯顿·M. 帕里斯　著
魏合义　刘学军　译

武汉大学出版社
WUHAN UNIVERSITY PRESS

图书在版编目(CIP)数据

城市环境生态学/(澳)柯尔斯顿·M.帕里斯著;魏合义,刘学军译.—武汉:武汉大学出版社,2021.11
书名原文:Ecology of Urban Environments
ISBN 978-7-307-22606-7

Ⅰ.城… Ⅱ.①柯… ②魏… ③刘… Ⅲ.城市环境—环境生态学 Ⅳ.X21

中国版本图书馆 CIP 数据核字(2021)第 195719 号

Ecology of Urban Environments, by Kirsten M. Parris, ISBN: 9781444332643.
Copyright © 2016 by John Wiley & Sons Ltd.
All Rights Reserved. This translation published under license. Authorized translation from the English language edition, Published by John Wiley & Sons. No part of this book may be reproduced in any form without the written permission of the original copyrights holder.
Copies of this book sold without a Wiley sticker on the cover are unauthorized and illegal.
本书中文简体版专有翻译出版权由 John Wiley & Sons, Inc.公司授予武汉大学出版社。未经许可,不得以任何手段和形式复制或抄袭本书内容。
本书封底贴有 Wiley 防伪标签,无标签者不得销售。

责任编辑:王 荣　　责任校对:李孟潇　　版式设计:马 佳

出版发行:武汉大学出版社　(430072　武昌　珞珈山)
　　　　　(电子邮箱:cbs22@whu.edu.cn　网址:www.wdp.whu.edu.cn)
印刷:武汉图物印刷有限公司
开本:720×1000　1/16　印张:17　字数:276 千字　插页:1
版次:2021 年 11 月第 1 版　　　2021 年 11 月第 1 次印刷
ISBN 978-7-307-22606-7　　　定价:49.00 元

版权所有,不得翻印;凡购我社的图书,如有质量问题,请与当地图书销售部门联系调换。

著者序

I am honoured to introduce the Chinese edition of *Ecology of Urban Environments*, and I am very happy that the book will now be accessible to a much wider readership. While designed as a text book for upper-level undergraduates and master students in ecology, environmental science, urban planning, architecture and landscape architecture, the book will also be a valuable resource for professionals involved in the construction and management of urban landscapes. The discipline of urban ecology and its integration with the built-environment professions have taken significant steps forward in recent years, but there is much more work to be done if we are to build sustainable, liveable cities—both for people and for nature. I hope that the publication of this book in Chinese will help further this goal.

Professor Kirsten Parris
Melbourne, Australia

译者序

在农村出生、成长并接受了基础教育，又几经辗转在城市学习、工作和生活，这些经历让我深切感受到中国农村的农业耕作和生活方式，也曾目睹过英国老牌现代城市的环境特点。随着专业知识和阅历的增加，"农村-城市"景象对比在感性上更强烈，农村向城市转变的规律与理性认识也逐渐深刻。

不可否认，近半个世纪我国在城镇化建设领域取得了巨大成绩，经济水平也突飞猛进。人们在为物质生活上取得的骄傲成绩而沾沾自喜时，却忽视了城镇化对生物、环境和人的消极影响。当前，城市环境中的热浪、雨洪、污染、废弃物、流性病、秩序紊乱、生物多样性、物质平衡以及能源消耗等问题，深深刺痛了城市管理者和居住者。

为了寻找应对策略、理论与方法，城乡规划学、风景园林学以及建筑学等领域中的研究者和实践者理应发挥积极作用。然而，在建筑类专业学生培养过程中，聚焦城市环境的生态学理论和生态观培养是缺失的。得益于本人负责双语教学的内在驱动，亟待引入一本前沿性、系统性、针对性以及适用性的教材。澳大利亚墨尔本大学 Parris 教授撰写的这本著作，完美符合这本理想教材的特征。考虑到我国多数高校仍然以中文教材为主，这本中文版《城市环境生态学》才凸显其价值所在。

专业图书翻译绝对不是语言好就能胜任。

为了确保此书的严谨性和可读性，我们对专业词汇进行多次校正，对一些生态学理论追根溯源，对整段语句"颠三倒四"。我们沮丧过，也彷徨过，但是，仍然坚持完成了此书的翻译。这项工作是正值全球新冠肺炎疫情暴发期间居家完成的，家人的陪伴给予我诸多精神鼓励。

此书的出版得到了国家自然科学基金（编号：31860233）、江西师范大学教改课题（编号：JXSDJG1812）以及城市建设学院教学经费资助，在此衷心感谢。

译　者

2021 年 8 月 20 日

序　　一

20世纪的前25年，我们见证了历史上的独特时代，人类成为地球演化的主要驱动力，正经历向人类世(Anthropocene)①的演变。这个快速增长的时期，是一个有大量人类活动并伴随对环境和生物过程产生极大影响的时代，被命名为"大加速"。该变化最清晰的指标是城市的出现，而城市则是人类的主要栖息地。在人类史的大部分时间里，大部分人是居住在乡村和广阔的原生态自然环境中，只有少数人居住在被称为"城市"的区域。1800年以来，快速的人口增长和工业化，随之而来的是城市的数量和规模的急剧增加，以及生活在城市区域的人口比例的增加。历史上第一次，人成为城市中的主要物种。

因此，从人类的视角来看，城市成为极端重要的环境：它是多数人居住的地方。它们构成了经济和文化活动中心，并且与创造它们的经济和文化一样多姿多彩。城市形态和规模各不相同，有的是从早期的贸易和航运中心发展而来的，有的则是在新区域发展起来。早期的城市布局趋于紧凑，市内交通适合步行和骑马。然而，廉价的汽车和公共交通已经使城市摆脱了早期的空间限制，现今许多城市向各个方向广泛蔓延。随着城市的发展，城市形态与功能的规划和管理日益重要，特别是城市对更有效和具有影响的私人建筑、公共空间和基本服务的需求增加。

然而，城市不只是人类的家园。多数城市是由建筑基础设施和开敞空间组成，如公园、花园、水域和自然残留斑块，这是城市的主要结构。这些空间栖息着大量的物种，其中有些茁壮成长，有些趋向灭绝。城市中的混合物种既包括该

---

①　"人类世(Anthropocene)"由诺贝尔化学奖得主保罗·克鲁岑于2000年提出，他认为人类活动对地球系统造成的各种影响将在未来很长一段时间内存在，未来甚至在5万年内人类仍然会是一个主要的地质推动力。

区域的土著物种，也有被引入并适应了城市环境的外来物种。就像城市提供人类活动的集中区域一样，它可以成为新生物发展的场所或物种间的新组合，也可以是物种间以新的方式与人类、建成环境的互动。

鉴于城市环境的重要性日益提高，以及城市中发展的生物系统的丰富性和魅力，而城市生态学才刚刚兴起似乎显得有点奇怪。的确，几十年前，生态学家似乎没有将城市视为想要研究的对象，而是寻找未受污染、生态"完好无损"的偏远区域。今天，城市已被看作一系列人造景观的终结，并且日益成为研究其生态功能的主题。城市中出现的物种如何生存和繁衍？生态群落如何在城市环境不断变化中发展？城市生态系统如何在水、营养和能量方面发挥作用？人类如何与这些环境相联系、改造并居住在其中？最后，是否有更好的方法规划和管理城市及其组成部分，从而为人类和各种生物群落带来更大的宜居性？

这就是城市生态学的内容，是一个不断发展的领域，旨在了解城市在生态方面以及人类和非人类栖居者之间是如何运作的。这本书的出版非常及时，它提供了综合的、引人入胜的城市环境生态学知识。在现有生态学理论的强大框架内，讨论了城市建设和扩张是如何影响城市环境的特征，以及人口、社区和生态系统的动态。本书也考虑了城市人口生态，并提供了在城市景观中保护生物多样性和维持生态系统服务的经典案例。总之，它旨在帮助我们更好地了解、规划和管理我们的主要栖息地，这一任务对于我们自己以及与我们共享城市的其他物种，在每一刻都变得更加紧迫和重要。

Richard J. Hobbs

西澳大学

序 二①

Telegraph Road

A long time ago came a man on a track
Walking thirty miles with a sack on his back
And he put down his load where he thought it was the best
Made a home in the wilderness

He built a cabin and a winter store
And he ploughed up the ground by the cold lake shore
The other travellers came walking down the track
And they never went further, no, they never went back

Then came the churches, then came the schools
Then came the lawyers, then came the rules
Then came the trains and the trucks with their loads
And the dirty old track was the Telegraph Road

Then came the mines, then came the ore
Then there was the hard times, then there was a war
Telegraph sang a song about the world outside

---

① Words and music by Mark Knopfler. Copyright (c) 1982 Chariscourt Ltd. International copyright secured all rights reserved. Reprinted by permission of Hal Leonard Corporation.

序 二

Telegraph Road got so deep and so wide
Like a rolling river

And my radio says tonight it's gonna freeze
People driving home from the factories
There's six lanes of traffic
Three lanes moving slow

I used to like to go to work but they shut it down
I got a right to go to work but there's no work here to be found
Yes, and they say we're gonna have to pay what's owed
We're gonna have to reap from some seed that's been sowed

And the birds up on the wires and the telegraph poles
They can always fly away from this rain and this cold
You can hear them singing out their telegraph code
All the way down the Telegraph Road

Mark Knopfler

致　　谢

将本书献给最支持我的 Mick 和 Owen，纪念城市生态学家 Joanne Ainley。

——Kirsten M. Parris

在此我要感谢很多人，没有他们对我的鼓励和支持，这本书仍旧没法完成。Alan Crowden 促使我考虑写作城市生态学教材；Mark Burgman 使我相信写这本书是个好主意。Alan 为这本书在 Wiley-Blackwell 出版提供了便利，从创作到完成，他和 Mark 是重要的支持者和指导者。

我非常感谢很多同事，感谢他们对该项目的兴趣以及对城市环境生态学作出的非常有帮助的探讨，主要有 Sarah Bekessy, Stefano Canessa, Jan Carey, Yung En Chee, Martin Cox, Danielle Dagenais, Jane Elith, Carolyn Enquist, Brian Enquist, Fiona Fidler, Tim Fletcher, Georgia Garrard, Leah Gerber, Gurutzeta Guillera-Arroita, Amy Hahs, Josh Hale, Andrew Hamer, Geoff Heard, Samantha Imberger, Claire Keely, Jesse Kurylo, José Lahoz-Monfort, Pia Lentini, Steve Livesley, Adrian Marshall, Mark McDonnell, Larry Meyer, Joslin Moore, Alejandra Morán-Ordóñez, Raoul Mulder, Emily Nicholson, Cathy Oke, Joanne Potts, Hugh Possingham, Dominique Potvin, Peter Rayner, Tracey Regan, John Sabo, Caragh Threlfall, Reid Tingley, Rodney van der Ree, Peter Vesk, Chris Walsh, Andrea White, Nick Williams 和 Brendan Wintle。

我衷心感谢我的家人、朋友和同事，在这本书的整个创作过程中获得了他们的支持。特别感谢 Michael McCarthy, Owen Parris, Ann Parris, John Gault, Bronwyn Parris, Monica Parris, Bridget Parris, Susan McCarthy, David McCarthy, Margery Priestley, Liz McCarthy, Tom McCarthy, Kirsty McCarthy, Sarah Bekessy,

致　谢

Michael Bode，Gerd Bossinger，Lyndal Borrell，Janine Campbell，Jan Carey，Jane Catford，Tasneem Chopra，Glenice Cook，Martin Cox，Kylie Crabbe，Karen Day，Jane Elith，Louisa Flander，Jane Furphy，Georgia Garrard，Cindy Hauser，Colin Hunter，Helen Kronberger，Rachel Kronberger，Min Laught，Sue Lee，Prema Lucas，Pavlina McMaster，Ruth Millard，John Moorey，Anne Macdonald，Meg Moorhouse，Sarah Niblock，Lisa Palmer，James Panichi，Rebecca Paton，Joanne Potts，Tracey Regan，Di Sandars，Anna Shanahan，Peter Vesk，Graham Vincent，Terry Walsh，Andrea White，Brendan Wintle 和 Ian Woodrow。

澳大利亚研究委员会，墨尔本大学理学院，澳大利亚城市生态学研究中心以及 NESP 清洁空气和城市景观中心为我的城市生态学研究提供了宝贵的支持。我要感谢我勇敢而热情的研究助手 Larry Meyer，他在该项目的许多方面为我提供了帮助。我非常感谢 Mark Burgman，Michael McCarthy，Caragh Threlfall，Chris Walsh 以及 2014 年研究生研讨会——墨尔本大学环境科学课程，在这研讨会上获得了非常有见解的评论。最后，我还要感谢在 Wiley-Blackwell 的 Ward Cooper，Delia Sandford，Kelvin Matthews，Emma Strickland，David McDade 和在 SPi Global 的 Kiruthika Balasubramanian，有了他们耐心的帮助，这本书才能顺利出版。

目　录

第1章　绪论 ·· 1
  1.1　场景设置 ··· 1
  1.2　什么是城市生态学？ ·· 1
  1.3　为什么城市生态学有价值？ ··· 3
    1.3.1　城市环境是广阔的和不断扩展的 ····························· 3
    1.3.2　城市环境的内在生态价值 ·· 6
    1.3.3　城市环境是检验和发展生态学理论的理想之选 ······· 7
    1.3.4　城市环境的性质影响着人类的健康和福祉 ··············· 7
    1.3.5　城市环境对于保护生物多样性至关重要 ·················· 8
  1.4　本书的目标 ·· 10

第2章　城市环境 ··· 18
  2.1　概述 ··· 18
  2.2　伴随城镇化的初级生物物理过程 ································ 23
    2.2.1　移除现有植被 ·· 24
    2.2.2　建筑、道路和其他城市基础设施的建设 ················ 25
    2.2.3　透水表面被取代 ·· 26
    2.2.4　开敞空间的减少 ·· 26
    2.2.5　水生生境的改变和破坏 ··· 27
    2.2.6　污染和废物的产生 ·· 29
  2.3　伴随城镇化的二级生物物理过程 ································ 32

目 录

  2.3.1 栖息地丧失、破碎化和孤立 ·· 32
  2.3.2 气候变化 ·· 32
  2.3.3 改变了水文环境 ··· 33
  2.3.4 空气、水和土壤污染 ·· 33
  2.3.5 改变了噪声和光环境 ··· 34
 2.4 城市环境的随机性 ··· 35
 2.5 总结 ··· 36

第 3 章 种群和物种水平对城镇化的响应 ··· 50
 3.1 概述 ··· 50
 3.2 响应城镇化的二级生物物理过程 ··· 52
  3.2.1 栖息地消失、破碎化和孤岛化 ·· 52
  3.2.2 气候变化 ·· 57
  3.2.3 改变了水文环境 ··· 58
  3.2.4 大气、水和土壤污染 ·· 59
  3.2.5 改变了噪声和光环境 ··· 62
 3.3 生物引进和入侵 ·· 67
  3.3.1 植物和真菌 ··· 67
  3.3.2 动物 ··· 69
 3.4 人为干扰 ·· 73
 3.5 对城市环境中种群的随机效应 ··· 75
 3.6 总结 ··· 76

第 4 章 群落响应城镇化 ·· 99
 4.1 概述 ··· 99
 4.2 选择：城市生态学中的生态位理论 ·· 102
  4.2.1 生态位和环境梯度 ·· 102
  4.2.2 栖息地模型 ··· 109
  4.2.3 生态组团和资源竞争模型 ·· 111

4.3	生态漂移：模拟城市群落中的随机性	115
4.4	扩散：个体在空间中的移动	117
4.5	多样性：城市环境中新谱系的进化	120
4.6	总结	122

第5章 生态系统响应城镇化 … 137

5.1	概述	137
5.2	碳	139
	5.2.1 碳循环的介绍	139
	5.2.2 城镇化对碳循环的影响	142
	5.2.3 缓解策略	145
5.3	水	149
	5.3.1 水循环介绍	149
	5.3.2 城镇化对水循环的影响	150
	5.3.3 缓解策略	152
5.4	氮循环	153
	5.4.1 城镇化对氮循环的影响	155
	5.4.2 缓解策略	157
5.5	总结	158

第6章 人类的城市生态学 … 174

6.1	概述	174
6.2	城市形态	178
	6.2.1 城市公园和开敞空间	178
	6.2.2 城市扩张和汽车依赖	183
	6.2.3 社区劣势和社区失调	186
6.3	污染和废物	187
	6.3.1 室外空气污染	189
	6.3.2 室内空气污染	190

6.4 城市环境中的气候变化 ··· 191
6.5 世界城市的卫生不平等 ··· 192
6.6 总结 ··· 194

第7章 保护城市中的生物多样性和维护生态系统服务 ························· 211
7.1 概述 ··· 211
7.2 城市生物多样性保护和维持生态系统服务的策略 ························· 214
　　7.2.1 将城市生态学与城市规划设计相结合 ································ 214
　　7.2.2 保护生物多样性景观特征和重要的生物物理资产 ··············· 216
　　7.2.3 发展绿色城市 ·· 220
　　7.2.4 维护或重建景观连通性 ··· 221
　　7.2.5 利用小空间 ·· 224
7.3 新型的栖息地和生态系统 ··· 227
7.4 总结 ··· 229

第8章 总结和展望 ··· 248
8.1 概述 ··· 248
8.2 我们是否需要一种新的城市生态学理论？ ··································· 248
　　8.2.1 城市生态系统的复杂性 ··· 249
　　8.2.2 人类对城市生态系统的控制 ·· 250
　　8.2.3 城市生态系统的独特性 ··· 250
8.3 城市生态学的定义和范围 ··· 251
8.4 我们需要一种新的城市科学理论吗？ ·· 252
8.5 未来方向 ·· 253

第1章 绪　　论

1.1　场景设置

当读这本书时,说明你居住在城市或城镇环境中,这是一个好的选择。如果你往窗外或门外看时,可以看到建筑、道路、汽车、围栏和街灯,以及人、猫、狗、树或花。你可以听到火车行驶在铁轨上的声音,撞击锤声,小提琴演奏声,孩子们的笑声或鸟鸣。你可能会闻到过往卡车的柴油机尾气味,附近餐厅里煮意大利面的香味,草坪新割后散发的清香,或者垃圾堆或明渠发出的恶臭。这就是城市生活的反差,在这里你可以找到人类生存最佳和最差的地方,人类为自己建造的栖息地可以补充或消除其他物种的栖息地。生态学家努力了解这个自然世界发展的过程和格局。直到最近,许多生态学家仍然在远离城市的地方实践生态科学,认为人类活动是对自然的破坏而不是自然的一部分。但是,生态原则也适应于城市环境,将人类与自然界剥离开来不利于我们研究城市生态。在21世纪高度城镇化的世界里,城市生态学是一门有重要价值的学科。

1.2　什么是城市生态学?

作为更广泛的生物学学科中的自然科学,生态学是研究生物的分布、多度、行为以及它们之间、它们与环境之间的科学。生态学跨越多个尺度,从个体生物内部,到个体、种群、群落和生态系统。生物是有生命的物质,如细菌、真菌、植物和动物。生态学家尚未将人类与其他生物一样作为生态学的一部分进行研究(人类行为生态学详见:Winterhalder, Smith, 2000; Borgerhoff Mulder, Schacht,

2012)。这是城市生态学与其他生态学之间的第一点区别。第二点是城市生态学的研究聚焦于城市环境，可以认为城市环境由人设计，为人服务。

在本书中，把城市生态学定义为城市环境中所有生物的生态学，包括在城市环境中的人类，以及受城市建设、扩张和城市运作影响的环境。例如：向城市人口提供饮用水的森林流域(集水区)。城市生态学包括人，因为人的存在、人口动态和人的行为，以及他们建造城镇时所发生的环境变化，对于我们了解城市系统是如何运作的至关重要。城市生态学在社会科学中具有不同的含义，它描述了一种利用生态学理论，解释城市结构和功能的城市社会学方法(见 Park, Burgess, 1967)。一些学者还将"城市生态学"描述为由自然科学、社会科学和人文科学融合在一起的交叉学科(如 Dooling et al., 2007；详见本书第 8 章对该观点的论述)。但是，本书的目的和焦点是致力于研究生态学的自然科学范畴。生态学本身是有益于城镇研究，而本书则提供一个广泛的综合概念，但是与其他城市生态学文献又有区别。与自然科学、社会科学和人文结合，可以促进人们理解城市生态学，为改善城市规划、设计和管理作出重要贡献，也造福于城市中所有物种。

城市生态学是一门相对年轻的学科，关于城市包含的内容，以及如何定义城市还存在一些争议(见 Collins et al., 2000；McIntyre et al., 2000；Pickett et al., 2001)。例如，我们是否应该通过城市中的人口数量或密度，一定特征的景观格局，建筑和道路的特征密度，或者这些地物要素的组合来确定(McIntyre et al., 2000；Luck, Wu, 2002；Hahs, McDonnell, 2006)？是每个人都应该使用的唯一城市定义，还是为适合于不同研究课题均作一个城市定义？Wittig(2009)支持"城市"这个词的狭义定义，城市中心区应由混凝土、沥青和建筑物，结合非原始的植被组成。这不包括城市的其他部分，如溪流、私人花园和残存自然植被区域。它也排除了城镇区域以外仍受其影响的环境。寻求在城市生态学研究中可以统一使用的"城市"定义是不现实的，因为随着研究规模和问题的提出，定义可能会发生变化。对于溪流或者猫头鹰来说，什么是城市？与对生活在城市中的人、甲虫或者真菌来说可能不太一样。但是，无论如何，定义必须清晰且定量，这样方法才能被借鉴，也能够与以往的相关研究相比较(McIntyre et al., 2000)。

1.3 为什么城市生态学有价值？

城市生态学的价值至少体现在以下五个方面：①城市环境是广阔的且不断扩展的；②生态学本身就有价值；③城市是检验和发展生态学理论的理想之地；④城市环境的状况会影响居民的健康和福祉；⑤城市也是保护生物多样性的重要场地。对城市生态学做进一步的研究不仅可以推动整个生态学科的发展，还将帮助我们拯救物种免于灭绝，维护生态系统功能和服务，以及改善人类健康，提高居民幸福感。特别是在人口增长和城镇化不断加快的时代，了解城市环境有助于我们创建更加宜居的城市，为人类和其他生物提供高质量的栖息地。下面将详细介绍这些论点。

1.3.1 城市环境是广阔的和不断扩展的

今天，全球一半以上的人居住在城市地区，这是有史以来第一次。自从工业革命以来，随着城市提供就业机会的增加、机械化程度的提高，以及对农业劳动力需求的下降，城市人口急剧增加。联合国人口基金(United Nations Population Fund，UNFPA)估计：到2030年全球城市人口数量将从目前的39亿增长到49亿，而到2050年将达到64亿[图1.1(a)]。相比之下，20世纪初的城市人口数量仅为2.2亿(UNFPA，2007；UN，2014)。仅仅过了130年，这个数值就增长22倍。发达国家的城镇人口数量将略有增长，而预期城镇居民数量增长较快的将出现在非洲、亚洲、拉丁美洲和加勒比海等发展中国家[图1.1(b)；UNFPA，2007]。城市生活转变所带来的社会和环境影响是深远的，但是不同地区之间也存在巨大的差异。

像澳大利亚和美国这些发达国家的城市扩张，一般通过在城镇郊区个别地块上建造房屋(图1.2)。大多数独栋房屋都分别由一个家庭居住，并且拥有电力管线、饮用水管道、一个或多个与封闭式排污系统相连的浴室、电话以及位于门前的硬化道路。一些房子拥有泳池，安装有空调。相对较大的土地面积只能容纳少数人，由此造成城市在此区域内的延展被称为城市扩张(Soule，2006)。相反，在撒哈拉以南的非洲、拉丁美洲和印度等地区的许多人定居于城市，被安置在城

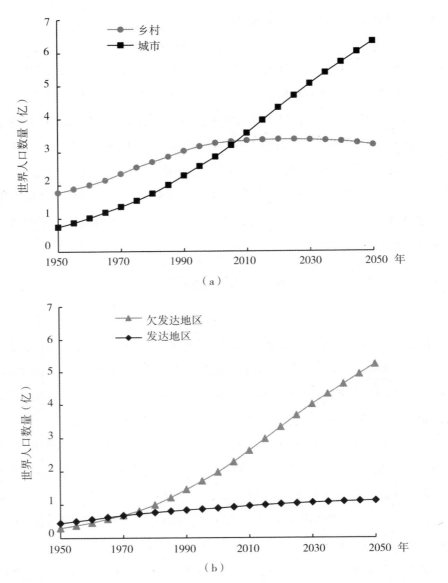

图 1.1 (a)世界城市和乡村人口;(b)欠发达地区和发达地区的城市人口(1950—2050 年)
[数据来源:联合国经济和社会事务部人口局(UN,DESA,2014)]

市内部或者边缘区域的非正规居住区(也称为贫民窟或棚户区)(UNFPA,2007)。这些临时居住地的人口密度高、卫生条件差,很少或者没有干净的饮用水,并且使用权没有保障(图1.3)。数百人可能共用一个浴室，饮用水可能被人类废弃物污染，住宅通常没有电力或者通风设备，而且没有铺路或者废弃物处理设施(Geyer et al.，2005；UNFPA，2007)。非正规居住区通常建在易遭受自然灾害的地方(如洪水和山体滑坡)，并且居住在此的人们没有合法权利。因此，非正规定居点的房屋很可能会在短期内拆除(Hardoy，Satterthwaite，1989；Tibaijuka，2005；Padhi，2007)。2005年，估计有10亿人(占世界总人口的1/6)生活在非正规居住区(UN-Habitat，2006)。

图1.2　美国内达华州拉斯维加斯的郊区(摄影：ulybug)
(https://creativecommons.org/licenses/by/2.0/.)

其他类型城市的发展则介于这两个极端之间。中密度和高密度的联排别墅和公寓是世界上许多城市的特征，它具有现代设施和基础设施，可提供高质量的生活标准，但其占用的空间比独立式住宅少。每种城市扩张的类型影响着新城镇化区域的生物多样性和生态系统功能(Liu et al.，2003)。例如，在澳大利亚某城市郊区用于建造大型低密度房屋并维持其居民生活方式的材料和能源，要比南非或

第 1 章 绪　　论

图 1.3　肯尼亚内罗毕的基贝拉贫民窟房屋（摄影：Colin Crowley）
（https：//creativecommons.org/licenses/by/2.0/.）

孟加拉国的一个棚户区的居民所使用的材料和能源多许多倍（Wackernagel，Rees，1996；McGranahan，Satterthwaite，2003）。居住在每种城市类型区域的人们所获得的健康和福利也有很大的差异（关于这点的讨论，请参阅本章后面内容）。因此，我们不能将城镇化（城镇建设）或城市扩张（城市人口增加）视为统一的过程。在未来几十年中，在发展中国家城市扩张将带来巨大的生态和社会挑战。笔者认为，提高对城市生态学的认识，可以更好地应对这些挑战。

1.3.2　城市环境的内在生态价值

城市环境具有内在生态价值，部分原因是它们与被替代的栖息地有很大的不同。生态系统、群落、物种和种群是如何适应生存环境从荒野、农业用地急剧变化为人类栖息地？哪些物种和群落能够繁衍生息？哪些遭受灭顶之灾？当本土物种消失而外来物种入侵时，会出现新的生物群落吗？如果是这样，即使它们在组成上有所不同，这些群落在功能上是否与它们替代的群落相似呢？城市区域功能可以充当城市生态系统吗？在特定的城镇化水平下，生态系统功能会崩溃吗？人类的偏好和行动，与城市生物多样性保护之间有什么关系？以及人的健康如何受空气污染、树木覆盖或开敞空间影响？我们已了解世界上某些地区的这些问题，但城市中更多的非人类生物、人和环境之间的关系，仍然需要进一步探讨。

1.3.3 城市环境是检验和发展生态学理论的理想之选

许多生态学理论，用于解释在相对原始的、未被人类干扰的生境中的生物扩散、多样性、行为和相互作用。例如：生态位理论（Hutchinson，1957）、种间竞争（Tansley，1917；Connell，1961）、最佳觅食理论（Charnov，1976）、捕食者与猎物的关系（Volterra，1926；Lotka，1932）、岛屿生物地理平衡理论（MacArthur，Wilson，1963，1967）、种群分布理论和斑块动力学（Levins，1969；Pickett，White，1985）、食物网（Hairston et al.，1960；Murdoch，1966）、集合群落理论（Gilpin，Hanski，1991；Leibold et al.，2004）、以及生物多样性和生物地理学中性理论（Hubbell，2001）。行为理论，如博弈理论（Maynard Smith，Price，1973），以及与动物交流、择偶和性选择理论（Zahavi，1975；Marten，Marler，1977；Wells，1977；Kirkpatrick，Ryan，1991），在很大程度上没有考虑城市环境中动物的行为，也已经发展起来了。

如 Collins 等（2000）所说，任何有价值的生态理论都应该适用于城市，以及郊区和荒野环境。一些理论已经在城市中进行了测试，例如：最佳觅食理论（Shochat et al.，2004）、生态位理论（Parris，Hazell，2005）、中间干扰假设（Blair，Launer，1997）、集合群落理论（Parris，2006）、多样性与生产力关系（Shochat et al.，2006）、食物网/营养动态（Faeth et al.，2005），这些理论都表现良好。这表明大部分现有的生态学理论（可能不是全部）也可以适用于城市区域。城市环境的动态性，也可以促进发展新的生态学理论，以及整合生态、社会和经济理论，以便于更好地研究城市系统的生态。

1.3.4 城市环境的性质影响着人类的健康和福祉

我们周围的环境显著而微妙地影响着人类的健康和福祉。在城市区域，对比最鲜明的是居住在安全、结构良好的房子里的人，与无家可归或者住在非正规居住区的人之间的差异。卫生设施不足，清洁饮用水缺乏以及极端天气条件下的防护不力，这些将会增加贫民窟的人患病的风险，而缺乏安全保障则使妇女更容易遭受暴力和性传播疾病，如艾滋病（Amuyunzu-Nyamongo et al.，2007；UNFPA，2007）。但是，城市环境的特征也会影响拥有住房居民的健康。城市中绿色自然和开敞的场所，为居民提供了锻炼身体和改善精神健康的机会（Giles-Corti et al.，2005；Gidlöf-Gunnarsson，Öhrström，2007）。最近的研究表明，城市扩张与肥胖率、交通死亡风险的增加有关（Ewing et al.，2003，2006，2016；Smith et al.，2008；Mackenbach et al.，2014）。不断扩展的社区内通常人行道较少，学校和商店等设施与居住区分开。因此，居民开车比骑行、步行的概率更大（Ewing et al.，2016）。

在最佳状态下，城市生活为社会互动、社区意识（社会资本）提供了机会，而这对于居民的福祉至关重要。然而，社会资本受到侵蚀，伴随高犯罪率、拥挤的生活条件；或者反之，社区不断延展导致社会孤立（Leyden，2003）。城市区域的高度社会失调，与居民患抑郁症风险的增加相关（Kim，2008）。城市中生物多样性和人类健康之间的关系，是有待进一步研究的兴趣点。最近的研究发现，随着稀树草原和其他绿色空间中生物多样性的增加，对游客的心理益处也随之增加（Fuller et al.，2007；Carrus et al.，2015）。

1.3.5 城市环境对于保护生物多样性至关重要

过去很多城镇都建在靠近河流、河口或者有庇护的区域，这些地方具有迷人的环境，也便于运输商品和人员往来。由于具有高生产力、相对温和的气候，以及在河流和海洋生境汇合的位置，这些区域的生物多样性通常也很高（Luck，2007）。如今，人口密度与生物多样性之间的关系仍旧是这样，物种丰富的地区仍然是人们优先定居之地（Cincotta et al.，2000；Luck et al.，2004；见专栏2.1）。例如，澳大利亚城市的发展是在大陆的沿海地区进行的，那里的降雨量、初级生产力和生物多样性都很高。由于剧烈的环境变化，城镇化经常引起人类需求和其他物种需求之间的冲突。

纵观各历史时期，城镇化可能导致了数千当地物种灭绝。McDonald 等

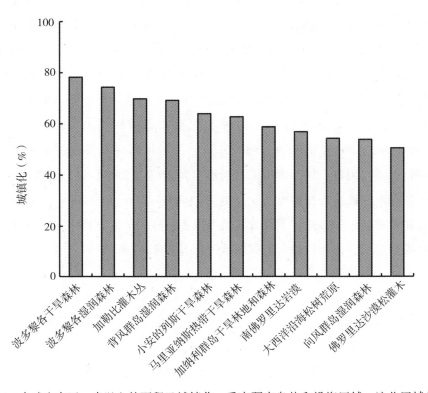

图1.4 全球生态区一半以上的面积已城镇化；重点研究岛屿和沿海区域。这些区域的当地物种被持续的城市扩张威胁（数据来源：McDonald et al.，2008）

(2008)估计 IUCN（世界自然保护联盟，Internation Union for Conservation of Nation）红色名录中的420种物种受到城镇化的威胁。当前，全球825个生态区中有11个区域的城镇化超过一半（图1.4），有29个区域的城镇化超过1/3（McDonald et al.，2008）。这29个生态区栖息着3056种物种，其中包括213种本土陆生脊椎动物，89种被列入IUCN红色名录。特定的功能群更有可能从市区消失，例如：陆生节肢动物、食虫鸟类、大型肉食动物、陆生脊椎动物，这些更容易受到外来掠食者侵害（Sewell，Catterall，1998；van der Ree，McCarthy，2005；Bond et al.，2006；Riley，2006）。在英国城市地区，矮小、喜阴植物对土壤水分要求较高，通常更容易消失。而喜爱开敞、干燥栖息地的高大植物则更容易繁衍（Thompson，McCarthy，2008；Duncan et al.，2011）。

虽然城镇化导致本土物种的消失，许多外来植物、动物被有意或无意地引入城市用于园林绿化，或者当作宠物饲养（McKinney，2002，2008；Tait et al.，2005）。一些物种对人类在城市环境中提供的资源表现出强烈、积极的响应，在城市中可以获得非常高的种群密度，这些被称为"共生物种"，或称为"城市适应者"（详见第7章）。但是，不要误认为生物保护在城镇中不重要，城市中心或边缘区域也有许多受威胁的本土物种和生态群落（见 Williams et al.，2005；Marchetti et al.，2006；Ives et al.，2016），特别是那些发育缓慢的物种（Vähä-Piikkiö et al.，2004）。持续城市扩张可能会危及这些物种和群落的生存，另外，还有一些离市区较远的物种也会受到威胁（McDonald et al.，2008），除非我们改变建造和管理城市的方式。不管城区是否位于自然保护区，将生态学理论和合理的城市规划相结合，可以最大限度地减少现有和未来城市区域生物多样性的消失（Luck et al.，2004）。

1.4　本书的目标

本书有两个目标。第一，通过综合现有的知识和已有的生态学理论，来确定复杂的城市生态系统的共性，从而使读者无障碍、前导性地了解城市生态学。第二，使城市生态学变得更有价值，使发达国家和发展中国家的学生、研究人员和政策制定者更关注城市生态学。迄今为止，许多城市生态学的研究都集中在北美和欧洲的富裕国家，或澳大利亚和新西兰。但是，如上所述，在未来几十年中，发展中国家的城市人口将大大增加，同时伴随一系列的社会和环境挑战。研究城市生态学对地球的未来发展是至关重要的，包括保护生物多样性、维持生态系统功能、保持社会凝聚力，以及提高人类健康和福祉。笔者希望这本书能够为世界各地的新兴城市生态学家提供信息，并启发他们的研究。

思考题

1. 你是如何定义"城市"一词的？可以考虑使用定性描述或定量指标。
2. 你对城市生态学哪个方面最感兴趣？为什么？
3. 描述城市中正规居住区和非正规居住区之间的差异。

4. 城市区域对保护生物多样性重要吗？用世界上至少3个城市的例子来论证你的观点。

5. 如何增加城镇中人与自然之间的联系？考虑多种不同的策略。

本章参考文献

Amuyunzu-Nyamongo M, Okeng'O L, Wagura A, et al. (2007) Putting on a brave face: The experiences of women living with HIV and AIDS in informal settlements of Nairobi, Kenya. *AIDS Care*, 19, 25-34.

Blair R B, Launer A E. (1997) Butterfly diversity and human land use: species assemblages along an urban gradient. *Biological Conservation*, 80, 113-125.

Bond J E, Beamer D A, Lamb T, et al. (2006) Combining genetic and geospatial analyses to infer population extinction in mygalomorph spiders endemic to the Los Angeles region. *Animal Conservation*, 9, 145-157.

Borgerhoff Mulder M, Schacht R. (2012) Human Behavioural Ecology. *eLS*, *Wiley*. doi: 10.1002/9780470015902.a0003671.pub2

Carrus G, Scopelliti M, Lafortezza R, et al. (2015) Go greener, feel better? The positive effects of biodiversity on the well-being of individuals visiting urban and peri-urban greens areas. *Landscape and Urban Planning*, 134, 221-228.

Charnov E L. (1976) Optimal foraging: the marginal value theorem. *Theoretical Population Biology*, 9, 129-136.

Cincotta R P, Wisnewski J, Engelman R. (2000) Human population in the biodiversity hotspots. *Nature*, 404, 990-992.

Collins J P, Kinzig A, Grimm N B, et al. (2000) A new urban ecology. *American Scientist*, 88, 416-425.

Connell J H. (1961) The influence of interspecific competition and others factors on the distribution of the barnacle Chthamalus stellatus. *Ecology*, 42, 710-723.

Dooling S, Graybill J, Greve A. (2007) Response to Young and Wolf: goal attainment in urban ecology research. *Urban Ecosystems*, 10, 339-347.

Duncan R P, Clemants S E, Corlett R T, et al. (2011) Plant traits and extinction in urban areas: a meta-analysis of 11 cities. *Global Ecology and Biogeography*, 20, 509-519.

Ewing R, Brownson R C, Berrigan D. (2006) Relationship between urban sprawl and weight of United States youth. *American Journal of Preventive Medicine*, 31, 464-474.

Ewing R, Hamidi S, Grace J B. (2016) Urban sprawl as a risk factor in motor vehicle crashes. *Urban Studies*, 53, 247-266.

Ewing R, Schieber R A, Zegeer C V. (2003) Urban sprawl as a risk factor in motor vehicle occupant and pedestrian fatalities. *American Journal of Public Health*, 9, 1541-1545.

Faeth S H, Warren P S, Shochat E, et al. (2005) Trophic dynamics in urban communities. *BioScience*, 55, 399-407.

Fuller R A, Irvine K N, Devine-Wright P, et al. (2007) Psychological benefits of greenspace increase with biodiversity. *Biology Letters*, 3, 390-394.

Geyer N, Mmuwe-Hlahane S, Shongwe-Magongo R G, et al. (2005) Contributing to the ICNP: validating the term "informal settlement". *International Nursing Review*, 52, 286-293.

Gidlöf-Gunnarsson A, Öhrström E. (2007) Noise and well-being in urban residential environments: The potential role of perceived availability to nearby green areas. *Landscape and Urban Planning*, 83, 115-126.

Giles-Corti B, Broomhall M, Knuiman M, et al. (2005) Increasing walking—How important is distance to, attractiveness, and size of public open space? *American Journal of Preventive Medicine*, 28, 169-176.

Gilpin M E, Hanski I A. (1991) *Metapopulation Dynamics: Empirical and Theoretical Investigations*. Academic Press, London.

Hahs A K, McDonnell M J. (2006) Selecting independent measures to quantify Melbourne's urban-rural gradient. *Landscape and Urban Planning*, 78, 435-448.

Hairston N G, Smith F E, Slobodkin L B. (1960) Community structure, population control, and competition. *American Naturalist*, 44, 421-425.

Hardoy E, Satterthwaite D. (1989) *Squatter Citizen. Life in the Urban Third World*. Earthscan Publications, London.

Hubbell S P. (2001) *The Unified Neutral Theory of Biodiversity and Biogeography*. Princeton University Press, Princeton.

Hutchinson G E. (1957) Concluding remarks. In *Cold Spring Harbour Symposium on Quantitative Biology* 22. Cold Spring Harbor, New York, 415-427.

Ives C D, Lentini P E, Threlfall C G, et al. (2016) Cities are hotspots for threatened species. *Global Ecology and Biogeography*, 25, 117-126.

Kim D. (2008) Blues from the neighborhood? Neighborhood characteristics and depression. *Epidemiologic Reviews* 30, 101-117. doi: 10.1093/epirev/mxn009.

Kirkpatrick M, Ryan M J. (1991) The paradox of the lek and the evolution of mating preferences. *Nature*, 350, 33-38.

Leibold M A, Holyoak M, Mouquet N, et al. (2004) The metacommunity concept: a frame-work for multi-scale community ecology. *Ecology Letters*, 7, 601-613.

Levins R. (1969) Some demographic and genetic consequences of environmental heterogeneity for biological control. *Bulletin of the Entomological Society of America*, 15, 237-240.

Leyden K M. (2003) Social capital and the built environment: the importance of walkable neighborhoods. *American Journal of Public Health*, 93, 1546-1551.

Liu J G, Daily G C, Ehrlich P R, et al. (2003) Effects of household dynamics on resource consumption and biodiversity. *Nature*, 421, 530-533.

Lotka A J. (1932) The growth of mixed populations: two species competing for a common food supply. *Journal of the Washington Academy of Sciences*, 22, 461-469.

Luck G W. (2007) The relationships between net primary productivity, human population density and species conservation. *Journal of Biogeography*, 34, 201-212.

Luck G W, Ricketts T H, Daily G C, et al. (2004) Alleviating spatial conflict between people and biodiversity. *Proceedings of the National Academy of Sciences of the United States of America*, 101, 182-186.

Luck M, Wu J. (2002) A gradient analysis of the landscape pattern of urbanization in the Phoenix metropolitan area of USA. *Landscape Ecology*, 17, 327-339.

MacArthur R H, Wilson E O. (1963) An equilibrium theory of insular biogeography. *Evolution*, 17, 373-387.

MacArthur R H, Wilson E O. (1967) *The Theory of Island Biogeography*. Princeton University Press, Princeton.

Mackerbach J D, Rutter H, Compernolle S, et al. (2014) Obesogenic environments: a systematic review of the association between the physical environment and adult weight status, the SPOTLIGHT project. *BMC Public Health*, 14, 1-15.

Marchetti M P, Lockwood J L, Light T. (2006) Effects of urbanization on California's fish diversity: Differentiation, homogenization and the influence of spatial scale. *Biological Conservation*, 127, 310-318.

Marten K, Marler P. (1977) Sound transmission and its significance for animal vocalization I. Temperate habitats. *Behavioral Ecology and Sociobiology*, 2, 271-290.

Maynard Smith J, Price G R. (1973) The logic of animal conflict. *Nature*, 246, 15-18.

McDonald R I, Kareiva P, Forman R T T. (2008) The implications of current and future urbanization for global protected areas and biodiversity conservation. *Biological Conservation*, 141, 1695-1703.

McDonnell M J, Pickett S T A. (1990) The study of ecosystem structure and function along urban-rural gradients: an unexploited opportunity for ecology. *Ecology*, 71, 1231-1237.

McGranahan G, Satterthwaite D. (2003) Urban centers: an assessment of sustainability. *Annual Review of Environment and Resources*, 28, 243-274.

McIntyre N E, Knowles-Yanez K, Hope D. (2000) Urban ecology as an interdisciplinary field: Differences in the use of "urban" between the social and natural sciences. *Urban Ecosys-tems*, 4, 5-24.

McKinney M L. (2002) Urbanization, biodiversity, and conservation. *BioScience*, 52, 883-890.

McKinney M L. (2008) Effects of urbanization on species richness: A review of plants and animals. *Urban Ecosystems*, 11, 161-176.

Murdoch W W. (1966) Community structure, population control and competition—a critique. *American Naturalist*, 100, 219-226.

Padhi R. (2007) Forced evictions and factory closures: rethinking citizenship and rights of working class women in Delhi. *Indian Journal of Gender Studies*, 14, 73-92.

Park R E, Burgess E W. (1967) *The City*. University of Chicago Press, Chicago.

Parris K M. (2006) Urban amphibian assemblages as metacommunities. *Journal of Animal Ecology*, 75, 757-764.

Parris K M, Hazell D L. (2005) Biotic effects of climate change in urban environments: the case of the grey-headed flying-fox (*Pteropus poliocephalus*) in Melbourne, Australia. *Biological Conservation*, 124, 267-276.

Pickett S T A, White P S. (1985) *The Ecology of Natural Disturbance as Patch Dynamics*. Academic Press, New York.

Pickett S T A, Cadenasso M L, Grove J M, et al. (2001) Urban ecological systems: Linking terrestrial ecological, physical, and socioeconomic components of metropolitan areas. *Annual Review of Ecology and Systematics*, 32, 127-157.

Riley S P. (2006) Spatial ecology of bobcats and gray foxes in urban and rural zones of a national park. *Journal of Wildlife Management*, 70, 1425-1435.

Sewell S R, Catterall C P. (1998) Bushland modification and styles of urban development: their effects on birds in southeast Queensland. *Wildlife Research*, 25, 41-63.

Shochat E, Lerman S, Katti M, et al. (2004) Linking optimal foraging behavior to bird community structure in an urban-desert landscape: Field experiments with artificial food patches. *American Naturalist*, 164, 232-243.

Shochat E, Warren P S, Faeth S H, et al. (2006) From patterns to emerging processes in mechanistic urban ecology. *Trends in Ecology and Evolution*, 21, 186-191.

Smith K R, Brown B B, Yamada I, et al. (2008) Walkability and body mass index. Density, design, and new diversity measures. *American Journal of Preventive*

Medicine, 35, 237-244.

Soule D C. (2006) *Urban Sprawl: A Comprehensive Reference Guide*. Greenwood Publishing Group, Westport.

Tait C J, Daniels C B, Hill R S. (2005) Changes in species assemblages within the Adelaide metropolitan area, Australia, 1836-2002. *Ecological Applications*, 15, 346-359.

Tansley A G. (1917) On competition between *Galium saxatile* L. (*G. hercynium* Weig.) and *Galium sylvestre* Poll. (*G. asperum* Schreb.) on different types of soil. *Journal of Ecology*, 5, 173-179.

Thompson K, McCarthy M A. (2008) Traits of British alien and native urban plants. *Journal of Ecology*, 96, 853-859.

Tibaijuka A K. (2005) *Report of the Fact-Finding Mission to Zimbabwe to assess the Scope and Impact of Operation Murambatsvina by the UN Special Envoy on Human Settlements Issues in Zimbabwe*. United Nations, New York.

UNFPA. (2007) *State of World Population* 2007. *Unleashing the Potential of Urban Growth*. United Nations Population Fund, New York.

United Nations, Department of Economic and Social Affairs, Population Division. (2014) *World Urbanization Prospects: The 2014 Revision*. United Nations, Geneva.

UN-Habitat. (2006) *State of the World's Cities* 2006/7: *The Millennium Development Goals and Urban Sustainability*. Earthscan Publications, London.

Vähä-Piikkiö I, Kurtto A, Hahkala V. (2004) Species number, historical elements and protection of threatened species in the flora of Helsinki, Finland. *Landscape and Urban Planning*, 68, 357-370.

van der Ree R, McCarthy M A. (2005) Inferring persistence of indigenous mammals in response to urbanization. *Animal Conservation*, 8, 309-319.

Volterra V. (1926) Variations and fluctuations of the numbers of individuals in animal species living together. Reprinted in Chapman R N. (1931) *Animal Ecology*. McGraw-Hill, New York.

Wackernagel M, Rees W. (1996) *Our Ecological Footprint: Reducing Human Impact on the Earth*. New Society Publishers, Gabriola Island, Canada.

Wells K D. (1977) The social behavior of anuran amphibians. *Animal Behaviour*, 25, 666-693.

Winterhalder B, Smith E A. (2000) Analyzing adaptive strategies: Human behavioral ecology at twenty-five. *Evolutionary Anthropology*, 9, 51-72.

Williams N S G, McDonnell M J, Seager E J. (2005) Factors influencing the loss of indigenous grasslands in an urbanising landscape: a case study from Melbourne, Australia. *Land-scape and Urban Planning*, 71, 35-49.

Wittig R. (2009) What is the main object of urban ecology? Determining demarcation using the example of research into urban flora. In *Ecology of Cities and Towns: A Comparative Approach*. McDonnell M J, Hahs A K, Breuste J eds. Cambridge University Press, Cambridge, 523-529.

Zahavi A. (1975) Mate selection: a selection for a handicap. *Journal of Theoretical Biology*, 53, 205-214.

第 2 章 城市环境

2.1 概述

在农业系统周围出现了人类定居点。大约在公元前9500年的中东地区从游牧、狩猎和采集生活方式转变为更加稳定的居住方式，这时期为新时期时代(Bellwood，2004)。到公元前3500年，第一批现代城市建立于美索不达米亚，也即今天的伊拉克地区。如今，人类居住区的规模从一片村庄到镇、城市和超过1000万人口的大都市(Pearce，1996)。2008年，全球有740个城市的人口数量超过50万，其中包括22个人口数量超过1000万的城市(Cox，2008)。当前，被归类为城市的面积比例，约占陆地表面的3%，并且这个比例还在上升(McGranahan et al.，2005；Seto et al.，2012)。然而，城市和城市居民对生态的影响远远超出了城市边界(Collins et al.，2000；Grimm et al.，2008a，2008b)。

在第1章，笔者简要讨论了人类历史上选择定居区域的类型，尽管人类倾向于选择生物多样性较高的沿海地区(Luck et al.，2004)，但是现代工程技术能够使人在原本荒凉的沙漠地区生活。例如，美国亚利桑那州的凤凰城大都市区生活着超过420万人，而阿拉伯联合酋长国的迪拜则有240万人(见专栏2.1)。千禧年生态系统评估(Millennium Ecosystem Assessment)认为，在六个生态系统中，沿海地区的城市土地覆盖率最高，人口密度也最高，其次是农耕区和内陆水域(栖息地临近河流和湖泊)(McGranahan et al.，2007)。Morris和Kingston(2002)提出一个理论，也即人类选择栖息地的目的是最大限度地提高个体健康。从历史上看，人类选择的这些区域能够获得充足食物和淡水，有安全庇护，并且不会受其他人群的威胁和野生动物的袭扰。该理论的假设是当人们意识到会获得更好的生活时,他们选择从一个地方(或栖息地)转移到另一个地方。

> 专栏 2.1

迪拜和凤凰城——两个沙漠城市的故事

迪拜

迪拜位于阿拉伯联合酋长国(图2.1)。人口从1975年的约18.3万增长到2015年的240万,在短短40年的时间里增长13倍(迪拜统计中心,2007,2015)。人口增长是由移民造成的,特别是外籍劳工移民以及自然增长推动(Pacione, 2005)。目前迪拜正经历着建设热潮,其中包括一些宏伟项目,如世界最高的建筑、世界上最大的购物中心、迪拜滨海项目(预计可以容纳120万人)以及一些形状如棕榈树和世界地图的离岸人工岛(Pacione,

图2.1 阿拉伯联合酋长国迪拜的摩天大楼(摄影:Jacqueline Schmid)
(https://pixabay.com/en/dubai-skyscraper-skyscrapers-639302/. Used under CC0 1.0 https://creativecommons.org/publicdomain/zero/1.0/deed.en.)

2005；UAE Interact，2007）。旅游业也蓬勃发展，2012 年的有 436 万游客来迪拜旅游（迪拜统计中心，2012）。2004 年，迪拜地理面积为 605km²，计划到 2015 年新增加 500km²（Pacione，2005）。到 2025 年，迪拜市区人口预计增长到 330 万（Cox，2008）。

迪拜位于阿拉伯沙漠，气候炎热干燥，年平均降雨量为 94mm，一年中有 6 个月的日平均最高温度超过 35℃（WMO，2015a）。迪拜没有河流，所以人消耗的水主要是淡化的海水。电力和水主要串联生产，天然气燃烧产生的废热被用来发电，也进行海水淡化（迪拜电力和水务局，2014）。2004 年，迪拜淡化水总消耗量为 243.2GL（1GL = 10^9L），2014 年增至 482.7GL（迪拜电力和水务局，2015a）。迪拜的快速发展也导致电力需求增加。电力消耗从 2004 年至 2014 年，增加超过了 2 倍，从 16.4GW·h 到 39.6GW·h（迪拜电力和水务局，2015b）。

凤凰城

凤凰城位于美国亚利桑那州（图 2.2）。凤凰城人口从 1950 年的 33 万增长到 2012 年的 420 万（Brazel et al.，2007；Guhathakurta，Gober，2007；Shrestha et al.，2012）。仅在 2004 年，在城市边缘和现有开发范围内，发放了超过 6 万份新房建造许可证（Brazel et al.，2007）。预计到 2030 年马里科帕县的人口（很大程度上由大城市凤凰城的人口决定）将达到 770 万（亚利桑那经济安全局，2006）。凤凰城迅速增长是由可利用水的增加、工业扩展以及人民涌入"太阳谷"寻求沙漠绿洲生活方式驱动的。

凤凰城位于索诺兰沙漠，气候干燥，一年中有 4 个月平均每日最高温度超过 35℃，年平均降雨量为 208mm（WMO，2015b）。该市的能源由火力发电、核电站、水力发电和一些太阳能电站供应（US EIA，2015）。人的饮用水及花园、游泳池、高尔夫场地和人工湖的用水，来源于萨尔特河、佛得角河和科罗拉多河上游集水区或地下蓄水层（Balling et al.，2008）。1995—2003 年，凤凰城的普通家庭每天用水量为 1788L，其中大部分用于室外设施，如园艺灌溉、泳池蓄水（Balling et al.，2008）。然而，由于采取了多种节水措施，近年来凤凰城的人均用水量一直在下降（Gammage et al.，2011）。

图 2.2 凤凰城市中心的天际线,照片经剪切和黑白转换(摄影:Sean Horan)
(https://creativecommons.org/licenses/by/2.0/.)

沙漠生活和生态足迹

世界自然基金会的地球生命力报告(WWF,2014),确定科威特、卡塔尔和阿拉伯联合酋长国为 2010 年世界人均生态足迹最大的三个国家。科威特的人均足迹为 10.4ghm^2(全球公顷),而美国的人均生态足迹为 7ghm^2(世界排名第 8 名),牙买加人均生态足迹为 2ghm^2(世界排名第 76)。一个国家的足迹是生产其消耗的资源(如燃料、食物、纤维和木材),吸纳其所生产废物以及为其基础设施提供空间所需的所有土地和渔场的总和(但是也要看到生态足迹概念的批评,Blomqvist et al.,2013)。2010 年,地球平均生物容量或总生产面积是 1.7ghm^2/人,其中,1 单位的全球公顷(ghm^2)是 1 公顷具有世界平均水平的资源生产和吸收废物能力的空间(WWF,2014)。2010 年,全球生态足迹为 2.6ghm^2/人,这表明人类的生活超出了地球的环境承载能力,我们需要 1.5 个地球才能维持当前的生态资源的需求。沙漠中的大城市,如迪拜和凤凰城,由于用于房子、汽车、办公楼和购物中心冷却系统的能源以及用于维护花园、高尔夫场地、水景和泳池的用水量,它们在生态足迹中所占的比例远高于国家的平均水平。城市建设使当地温度增加,高于周围地区温度,这种现象称为城市热岛效应(见专栏 2.2)。随着城市变得越来越大和密集,它们也会变得越来越暖。因此,在炎热的沙漠城市,需要更多的能源冷却室内和汽车内,也需要更多的水用于维护花园、庭院和泳池(Baker et al.,2002;Guhathakurta,Gober,2007;Grimm et al.,2008a)。

城市生活可以提供适当的住所,清洁饮用水和医疗保健。因此,城市人的健康程度要比乡村高(Van de Poel et al.,2009;WHO,UN-Habitat,2010;World Bank,2013)。但是,生活在城市贫民窟的人(明显比生活在规划良好的城市区域的人更差),有可能比生活在乡村的人的健康状况还要差,一些疾病患病率、营养不良和婴儿死亡率也更高一些(Dyson,2003;UNFPA,2007;WHO,UN-Habitat,2010;World Bank,2013)。毫无疑问,许多人从乡村移居到城市,是想要改善他们的生活环境,如果目标没有实现,他们是否愿意再次返回?这个值得商榷。

城市建设,城市增长的步伐和发展形势(如城市扩展、高密度住房、非正规居住区)受经济、社会和文化等因素的相互影响,并随时间推移而不断变化(Lyon,Driskell,2011;Macionis,Parrillo,2012)。与城镇化的结果相比,本书较少关注社会经济对城镇化的驱动(见 Henry et al.,2003;Alig et al.,2004;Liu et al.,2005),而种群、物种、生态群落、生态系统功能以及城市居住者的健康和福祉是本书的研究重点。本章主要探讨了城镇化过程,城市栖息地的特征,以及其与乡村或非城镇栖息地的区别等。

专栏 2.2

城市热岛效应

城市热岛效应最早是由 19 世纪初,业余气象学家 Luke Howard 提出的,根据他的观察,伦敦市中心的夜间最低气温高于周围乡村的气温(Howard,1833)。热岛之所以在城市地区形成,是因为用于建筑物、道路和人行道使用的材料比天然植被吸收并保留更多的太阳辐射能，白天吸收的热量随后在晚上释放到大气中。工业、车辆、建筑的供热和制冷产生的热量，以及城市中颗粒物空气污染和风速的降低均是产生城市热岛的因素（Gartland，2008）。植被被不透水路面替代，也减少了因蒸发而损失的热量，从而进一步增加了白天的热储量和夜晚的热量释放（Gartland，2008）。Brazel 等（2007）发现，在 1990—2004 年，亚利桑那州凤凰城的气象站半径 1km 范围内，随着 1000 座新房建设完成，6 月份平均最低温度提高了 1.4℃。在中国北京，不透水面的比例较高导致约 80% 的夏季地表温度的局部空间变化（Ouyang et al.，2008）。

在温带气候区，在冬季晴朗、无风的夜晚，热岛效应最显著。例如，在 1992 年 8 月的一个晴朗夜晚，穿过澳大利亚墨尔本的温度横断面显示，与城市西边界的乡村地区相比，中央商务区的最高升温效应为 7.1℃（Torok et al.，2001）。在热带的新加坡地区，也观察到类似的现象，一年中较为干燥的月份中，晴朗、无风的夜晚，城市热岛效应最为明显（Chow，Roth，2006）。城市热岛效应可以实质性地升高夏季午后和夜间的温度，导致炎热气候中人类不适、患病和死亡（Brazel et al.，2000；Harlan et al.，2007；Tan et al.，2010）。城市热岛可以通过多种方式缓解，例如使用冷却屋顶和铺装材料，以及增加植被覆盖率，如屋顶绿化和垂直绿化方式（Gartland，2008；Gago et al.，2013）。通过这种方式为城市降温，可以为居住在那里的人们带来一些益处，包括节能、清洁的空气、提高健康和福祉（Gartland，2008）。增加植被覆盖，也有益于当地的生物多样性。

2.2 伴随城镇化的初级生物物理过程

尽管全球不同地区的地质、植被和气候存在差异，但是与城镇建设相关的生物物理过程通常相似。城市栖息地也具有许多自然特征，为非人类物种的生存带来机遇和挑战。城镇化的初级生物物理过程，包括移除现有植被，建筑、道路、照明系统、排水系统、围栏和其他城市基础设施建设，不透水路面替代透水性表面，改变或破坏水生生境，如池塘、溪流和河流，以及污染和废物的产生（表2.1）。下面将详细讨论每一项内容。

表2.1 城镇化的初级生物物理过程以及由此产生的二级生物物理工程

	初级过程	二级过程
1	移除现存植被	栖息地消失，破碎化和孤岛化（来源于1,2,4,5）
2	建筑、道路、照明、排水管和其他城市设施建设	气候变化（来源于1,2,3,4）
3	不透水表面取代透水表面	改变了噪声环境（来源于1,2,3）
4	减少了开敞空间	改变了光环境（来源于2,3）
5	改变或破坏了水生生物栖息地	改变了水文环境（来源于1,2,3,4,5）
6	污染和废弃物产生	空气、水和土壤污染（来源于1,2,3,6）

2.2.1 移除现有植被

城镇建设过程中，原有植物被大量清除。根据城镇化区域的气候、土壤类型和土地使用历史，现存的植被可以是本土的原始森林、林地、荒地、草原或稀疏的沙漠植被，次生或再生植被，杂草或牧场草。一般而言，城镇的植被覆盖率随着人口密度或城区建筑密度的增加而下降（Pauchard et al., 2006; Kromroy et al., 2007）。值得注意的是，城市密度（Urban Density）已有多种定义和衡量方式，具体取决于研究者所关注的本质和问题，例如：单位面积的人口、住宅和就业数量；单位面积的建筑面积或开敞空间（Dovey, Pafka, 2014）。研究美国在天然森林区域建造的12个城市，发现平均林木覆盖率最高的是公园和荒地，居民区域为中等，而商业和工业区则最低（Nowak et al., 1996；图2.3）。

Er等（2005）推测，在1859—1999年，加拿大温哥华的87%的森林面积已经被城市建设取代。有趣的是，城市扩展可以增加非城镇地区的本土植被的净覆盖率。例如，在1991—2000年，波多黎各岛上的森林覆盖率从28%增至40%，在放弃农业生产（撂荒）的同时，村民从乡村向城市区域转移（Parés-Ramos et al., 2008）。随着城镇化的发展，公园中和街道树木的种植可以增加城市林木覆盖率，部分补偿了城镇化过程中植被的最初损失。但是，通常这些种植的物种不是当地物种，也无法取代本土的植被群落(Tait et al., 2005; Shukuroglou, McCarthy, 2006)。

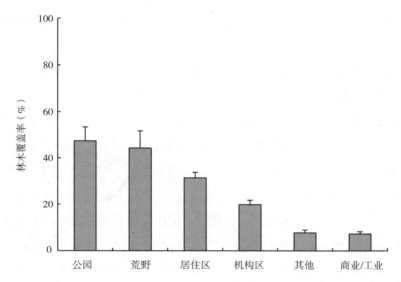

图2.3　美国建在原始森林区域的12个城市，不同土地使用类型中林木覆盖率(%)。荒野地=空地或荒地，机构=机构区域如医院或学校，其他=农业土地、果园、道路和机场，商业/工业=商业和工业用地，误差线代表标准误差（数据来源：Nowak et al., 1996）

2.2.2　建筑、道路和其他城市基础设施的建设

城镇化过程中最明显的物理过程，可能是住宅和其他建筑物的建设，以及其他辅助性设施的建造，例如：围墙，铺设的道路和人行道，水电供应设施，雨水排水系统，下水道和路灯。正规的居住地往往建有结构合理的住房和完善的基础设施。相反，非正规的居住区通常由临时住宅组成，没有电力、卫生设施，没有干净的水和规划很好的道路。许多城市的中央商务区(CBD，也称为市中心或城市核心)，由密集的办公楼、公寓楼、商店、酒店、法院和宗教建筑组成。住宅区(郊区)倾向于设计较低的建筑物密度(McKinney, 2002)，特别是在有大型住宅区的富裕社区。

2.2.3 透水表面被取代

透水表面是指允许水渗透的表面，而不透水表面是指不允许水渗透的表面。城市建设时，由于土壤、沙粒、树木被建筑物和铺砌的地面取代，不透水面积大大增加(McKinney，2002)。在城市中，降雨时大部分雨水降落到房屋和其他建筑物上，柏油路、停车场或混凝土车道、人行道上。在多数情况下，雨水排水系统将雨水直接汇流到运河、溪流或海洋中(Walsh et al.，2004)。另外，除了阻碍雨水的下渗外，建筑物、沥青和混凝土还吸收了比自然表面更多的太阳光短波辐射，并以热量的形式释放出来(Bridgman et al.，1995；见专栏 2.2)。与土壤和植被相比，不透水表面可以增加声音的反射率(Warren et al.，2006；Slabbekoorn et al.，2007)。因此，不透水表面比例极高的区域，相比其他区域会更热和嘈杂。最后，铺砌的道路阻断了空气和水分(Scalenghe，Marsan，2008)，干扰了生物过程，如营养物质循环和气体交换，并且阻止了植物根系或挖穴动物穿透地面。

Scalenghe 和 Marsan(2008)推算，欧洲9%的地面被不透水材料覆盖，而全球不透水表面可能超过 $50\times10^4 km^2$，或占 0.34%的全球陆地面积(Elvidge et al.，2007)。在城市内部，不透水表面的比例通常随城市密度的增加而增大。例如，最近的一项研究发现，中国北京市的六个同心区的不透水表面的平均覆盖率，从市中心的 67.3% 到最外围的 9.3% 不等(Xiao et al.，2007)。在 $4084km^2$ 的研究区域中，不透水表面的平均覆盖率为 20.8%，但是在城市核心区的不透水表面覆盖率达 90%(Xiao et al.，2007)。调查显示美国 37 个城市中，商业和工业社区的不透水表面覆盖率最高，在居住区较低，而在公园和荒地的不透水表面覆盖率最低(Nowak et al.，1996)。

2.2.4 开敞空间的减少

开敞空间(也即绿色空间或绿色开敞空间)具有较少的建筑物或道路，包括城市公园、自然保护区、湿地、河道、水库、市场花园、机场、运动场、高尔夫场地、荒地和废弃地(vacant lots)。开敞空间随着建筑物、道路和其他城市基础设施的建设而下降，并且与不透水表面的覆盖率成反比(见 American Forests，2003；图 2.4)。城市中的公共开敞空间为人类提供了娱乐休闲场地，以及动植

2.2 伴随城镇化的初级生物物理过程

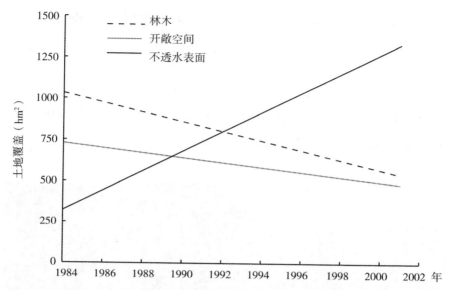

图 2.4　1984—2001 年，城市扩张导致美国北卡罗来纳州科尼利厄斯市的土地覆盖变化：不透水表面增加，而林木和开敞空间覆盖率减少（数据来源：American Forests，2003）

物的栖息地。但是，随着城市人口的增加，开敞空间常常因城市建设而减少。例如，在1990—2000年，美国连续减少了 $1.4 \times 10^4 \text{km}^2$ 的城市开敞空间，每个城市开敞空间减少的面积与人口增长密切相关（$r^2 = 0.72$；McDonald et al.，2010）。物种-面积的关系意味着，更大的开敞空间通常可以支撑更高的物种多样性，包括具有相对复杂的生活史、活动面积需求大以及对人为干扰或噪声敏感的动物（Fernández-Juricic，2000；Angold et al.，2006；Murgui，2007）。然而，对全球 38 个城市区域的调查显示，只有 16 个城市区域的开敞空间大于 4km^2（Forman，2008）。在 11 个城市地区中，最大开敞空间面积小于 1km^2。但是，即使很小的开敞空间区域，也可以为城市本地动植物的保护作出贡献（Gibb，Hochuli，2002；Rosensweig，2003；Fuller et al.，2007），并为人类健康和福祉提供重要支持（Matsuoka，Kaplan，2008）。

2.2.5　水生生境的改变和破坏

城市建设过程中，许多水生生境被改变或者消失：池塘和沼泽被填充或者紧密限制在工程围墙内;溪流和河流被取直,使用堤坝和水闸控制着,转入地下,或者使用混凝土铺砌河道;季节性湿地和滩涂地被设计整齐的公园替代(图2.5)。这些措施中的一部分旨在控制洪水,或者消除携带病原体的蚊子的繁殖场地(Meredith,2005)。湿地经常被破坏,并为城市发展让路(Kentula et al.,2004;Pauchard et al.,2006)。例如,在1975—2000年,城镇化破坏了智利的康塞普西翁的1734hm²(23%)湿地(Pauchard et al.,2006)。这些湿地中的大多数被抽干和填充,为住宅和工业发展提供了空间。

图2.5 英国约克郡的Ouse河狭窄的河道(摄影:Kirsten M. Parris)

因水文条件或土地管理政策的变化,城市环境中残存的湿地的结构也发生了改变。管理者可能会修剪蓬乱的沼泽或者河岸地区以及周围植被,使它们显得整洁且维护良好(Nassauer,2004)。除了人为改变城市水生生境之外,城镇化也导致其发生间接的改变。这些通常包括河流中悬浮沉积物增加,河岸的侵蚀和河道的扩大(Suren et al.,2005;Chin,2006)。在一些城市的修复项目,试图将水道和湿地恢复其自然状态,以造福人类和其他生物。这些项目包括河流廊道和湿地的植被恢复，以及用天然堤岸代替混凝土河道的实验（见 Hynes et al.，2004；Suren et al.，2005）。

2.2.6 污染和废物的产生

人类和工业生产排放的污染物和废弃物，集中在人口密集居住、旅行和工作地。通常，它们包括人为产生的废弃物，家庭和工业垃圾，工业和汽车尾气造成的大气污染，家庭和工业废弃化学品，以及氮和磷等营养物质。克里特岛的米诺斯文明时期的城市（公元前 3000—公元前 1100 年）和位于今天的巴基斯坦、印度和阿富汗的哈拉帕/印度河流文明（公元前 2600—公元前 1900 年），拥有第一个已知的城市下水道和卫生系统，包括冲洗厕所，与砖砌下水道相连的每个房屋的排水系统，房屋墙壁间建有垃圾槽，以及公共垃圾桶（Angelakis et al.，2014；Khan，2014）。今天，城市之间污染物和废弃物管理系统的差异很大，这取决于基础设施的水平，环境保护法规和服从性，以及非正规居住区的程度。

有效地收集、运输和处理人类废弃物对于城市居民的健康至关重要，特别是在人口密度高的地区（McMichael，2000）。发达国家的城市往往比发展中国家的城市拥有更多的下水道系统和更加先进的污水处理厂（Harada et al.，2008；Beyene et al.，2009），以及更强的对大气污染的产生、家庭和工业废弃物的处理能力的立法控制（Pargal et al.，1997；Hettige et al.，1998；UNEP，2005）。不同国家、不同城市之间，大气污染物（例如颗粒物）的浓度差异也很大（图 2.6）。例如，2010 年，印度新德里 PM_{10} 的年平均浓度为 $286\mu g \cdot m^{-3}$，玻利维亚拉巴斯的 PM_{10} 的年平均浓度为 $42\mu g \cdot m^{-3}$（WHO，2014a）。世界卫生组织有关 PM_{10} 的空气质量指南是年平均值 $20\mu g \cdot m^{-3}$（WHO，2014b）。

2004 年，全世界范围内将有 $1.3Pg(1.3\times10^{15}g)$ 的城市垃圾（家庭垃圾）产生，其中还不包括来自非经合组织国家的印度、中国等乡村地区的垃圾（Lacoste，Chalmin，2007）。到 2025 年，预计将产生 2.2Pg 的城市垃圾（Hoornweg，Bhada-Tata，2012）。美国是世界人均废弃物产生率最高的国家（Lacoste，Chalmin，2007），2012 年每人平均产生 725kg 城市固体废物（US EPA，2014；图 2.7(a)）。有趣的是，自 1990 年以来，美国人均废弃物产量一直保持不变或下降，但是这些废弃物的可回收利用或堆肥比例从 16.0% 增至 34.5%，并减少了垃圾填埋场的

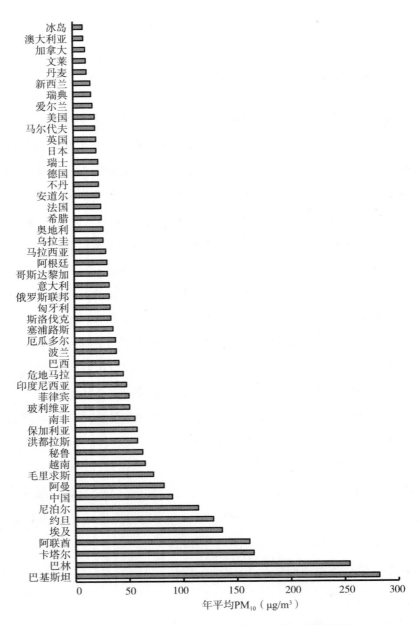

图 2.6　50 个国家的大气中年平均粗颗粒（PM_{10}）浓度（$\mu g/m^3$）（数据来源：WHO，2014）

数量（US EPA，2014；图 2.7（b））。在肯尼亚内罗毕，每位居民平均产生 183kg 的城市固体废物（UNEP，2005）。在城市贫民窟和非正规居住区，几乎没有正式

2.2 伴随城镇化的初级生物物理过程

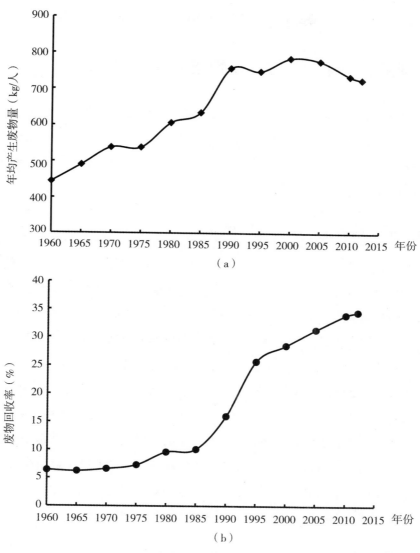

图2.7 1960—2012年美国城市的固体废物：(a)年均产生废物量(kg/人)；
(b)废物回收率(数据来源：US EPA，2014)

的废物收集服务。这导致的结果是，内罗毕存在严重的乱丢垃圾和乱倒垃圾所造成的污染问题（UNEP，2005）。不可生物降解的塑料袋尤其是典型问题，因为它们会堵塞排水沟和水渠，为携带疟原虫的蚊子提供繁殖的栖息地，并降低农田的生产力(Njeru, 2006)。2007年, 肯尼亚政府禁止使用薄塑料袋, 并对较厚的塑料袋征收高额税款, 以减少塑料袋的使用(BBC News, 2007-06-14)。

2.3 伴随城镇化的二级生物物理过程

城镇化过程中, 初级生物物理过程会导致一系列的二级生物物理过程, 这些过程对城市生态环境产生了深远的影响。其中包括生境的丧失、生境破碎化和孤立, 气候改变、水文和光照环境改变, 增加了噪声水平以及空气、水和土壤的污染(表2.1)。

2.3.1 栖息地丧失、破碎化和孤立

如上所述, 城市建设时, 陆地和水生的生物栖息地都将消失。很多情况下, 残余的栖息地将被分割成小块区域或小片, 这些地方往往被城市基础设施、人和快速行驶的车辆隔离。随着时间的流逝, 城市环境中残留的小片栖息地将进一步发生变化。它们的周长面积比高, 意味着它们更容易受到边缘效应的影响, 如光线变化、热力和风力状况, 杂草的入侵、人类的踩踏(Cilliers et al., 2008; Hamberg et al., 2008)。城市湿地也很容易受到外来动物的入侵(Spinks et al., 2003; Riley et al., 2005; Bury, 2008), 如鱼类和乌龟等外来动物。一般而言, 城市景观比乡村景观更加破碎化, 在相对较小的区域内, 会出现许多土地利用类型或小块的生物群落(Luck, Wu, 2002; Breuste et al., 2008)。群落生境是欧洲生态学家使用的一个术语, 用于描述与特定生态群落相关, 相对统一的环境条件(Löfvenhaft et al., 2002, 2004)。

2.3.2 气候变化

城市建设会引起明显的气候变化。尽管其中最典型的是城市热岛效应(见专栏2.2), 但是城镇化也会增加云量, 减少太阳辐射、改变降雨、湿度和风力, 并增加雷暴活动(Bridgman et al., 1995; Changnon, 2001; Sturman, Tapper, 2006)。制造业、发电、蒸发冷却以及公园、花园灌溉等人类活动, 将水分释放到大气中, 增加了城市的绝对湿度(但需要注意的是, 相对湿度可能较低, 因为

空气更热，可以容纳更多的水分)(Bridgman et al.，1995)。在静止状态下，湿度增加和大气中颗粒污染物(提供凝结核)的结合会促进雾的产生。在冬季，城市雾天数量可能是乡村地区的两倍多(Landsberg，1981)。在有利于空气垂直对流的条件下，城市中的大气湿度、空气污染、热量、风速降低，会增加云的形成(Bridgman et al.，1995)。尽管在没有云的情况下，空气中的颗粒污染物也会减少太阳辐射，但是这会增加雾霾度和降雨量，并减少太阳辐射量(Sturman，Tapper，2006)。在有风的条件下，不同高度的城市建筑物(或粗糙度)可以降低风速，低于乡村地区的风速，但是高层建筑也可以产生风漏斗和局部的"风洞"效应(Bridgman et al.，1995)。

2.3.3 改变了水文环境

将城市集水区的不渗透表面改为渗透材料，以及建设高效的排水系统，在很大程度上可改变城市的水文环境。落在不透水屋顶和地面上的雨水，通过排水系统迅速流到溪流中，导致溪流水量突然增加，随后迅速下降(Walsh et al.，2005;图 2.8)。相反，雨水渗透减少，会降低土壤含水量和地下水的流量，这可能使城市河流的基流量降低到乡村集水区河流基流量以下。因此，与乡村河流相比，城市河流的基流量较低，高峰流量却较高，此外还会发生暴洪事件(图 2.8)。这些短暂、高流量的暴洪往往会侵蚀堤岸、冲刷沟渠以及降低结构的复杂性(Walsh et al.，2005)。河道水流方式和栖息地的变化，对河流生物群，包括无脊椎动物、鱼类、两栖动物和爬行动物，均产生重要影响。

2.3.4 空气、水和土壤污染

工业排放废气、汽车尾气和固体燃料燃烧通常是城市环境中最明显的空气污染源。但是，污染的土壤、沉积物和排水沟也很常见。根据污染物(如柴油微粒、一氧化碳、臭氧、苯、重金属、氮、磷、农药和除草剂残留物，城市废物或人类废物)的类型和浓度，城市污染会损害人、动物和其他植物的健康。例如：暴露在污染的空气中，人会增加罹患心血管疾病、呼吸系统疾病(如哮喘、肺气肿和癌症)以及不育的风险(Brook，2008;Krivoshto et al.，2008)。草甘膦除草剂，通常在城市和乡村地区使用，也可以杀死幼虫期、成虫期的两栖动物(Relyea,

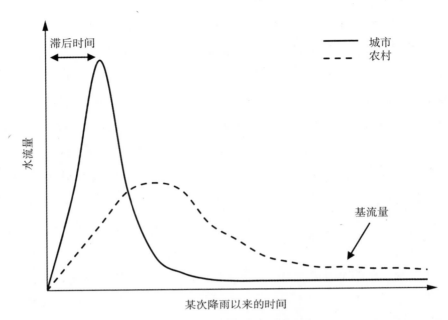

图 2.8　城市和乡村地区自降雨后溪流水流量与时间的关系。与乡村曲线相比，城市曲线的滞后时间更短、峰值更高、沉降更快、基流量更低，这些特征反映了城市河流的"瞬间性"

2005）。城市中心下风处大气中的氮沉积，对有些物种有利，但也会损害其他物种，因此改变了城市植物和微生物群落结构（Fenn et al.，2003）。在美国西部，土壤含氮量高，已增强了外来草对植物群落的入侵，而外来草则与本地草本植物和濒危蝴蝶的寄主植物竞争（Fenn et al.，2003）。城市垃圾（家庭垃圾）为世界各地城市的多种动物提供了食物，无论是本地动物还是外来动物（参见 Beckmann，Berger，2003）。

2.3.5　改变了噪声和光环境

通常城市比乡村地区更嘈杂，但是城镇化也在其他方面改变了噪声环境。城市噪声的频率分布与乡村地区不同，一天中噪声的时间和持续性，以及通过复杂城市景观结构中的回音和噪声衰减也有差异（Patricelli，Blickley，2006；Warren et al.，2006）。陆地城市的噪声大部分来自道路上的车辆通行，其他来源像工业生产、建设、过往的飞机和火车、割草机、吹叶机、放大的音乐声，这些都可能造成噪声(Warren et al., 2006)。在河流、湖泊、河口、港口等水生生境中,船只交通以及桥梁、堤岸建设产生的水下噪声可能很大(Lesage et al., 1999; Haviland-Howell et al., 2007; Graham, Cooke, 2008; Bailey et al., 2010)。城市也比乡村地区更亮一些。当人工灯光改变了自然光的昼夜循环,就会发生生态光污染(Longcore, Rich, 2004; Gaston et al., 2013)。世界各城市的生态光污染非常显著,可以从太空中清楚地看到(图2.9)。

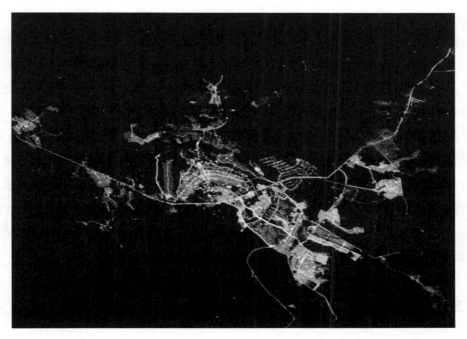

图2.9 夜晚从太空看巴西,巴西利亚的生态光污染(图片由美国NASA约翰逊太空中心地球科学和遥感组提供)

2.4 城市环境的随机性

在任何栖息地中,环境条件如温度、降水、风、土壤湿度水平和食物可用性以及其他资源,可能在每天、每个季节以及每年都不相同。环境随机性可以定义为环境随时间的中、小尺度变化,这些变化会影响种群个体的出生率和死亡率

(May，1974；Lande，1993)。它通常需要与种群随机性区别开来，种群随机性是由于种群中偶然的出生和死亡(大灾难)造成的，以及大规模的环境扰动导致种群剧烈下降(Lande，1993)。在第3章中，我们将讨论不同形式的随机性对种群的影响，但是这里值得考虑的是，与非城市环境相比，城市环境可能会经历或多或少的环境随机性。

伴随城镇化过程的各种生物物理过程，并没有明确的方式，如2.3.1小节所述，城市栖息地通常具有空间变异性或斑块特征(Niemela，1999；Fernandez-Juricic，Jokimaki，2001；Luck，Wu，2002)。但这并不一定意味着城市中某一特定位置随时间高度变化。人类对城市栖息地的频繁干扰，可能会导致特定物种经历的环境条件随时间变化程度增加(Rebele，1994；Sattler et al.，2010)，如草本、节肢动物。但与邻近的乡村地区相比，城市中的某些环境条件实际上变化程度可能较小。与城市热岛效应以及公园、庭院人工浇水有关的城市气候变化，会减小空气中温度、湿度和土壤湿度的昼夜、季节性变化(Oke，1982；Parris，Hazell，2005)。此外，对于杂食鸟类、果食性蝙蝠和食腐哺乳动物等广食性物种的食物供应，城市地区的季节变化可能远低于乡村地区，若全年都有充足的食物供应，种群密度会很高(Beckmann，Berger，2003；Contesse et al.，2004；Shochat，2004；Williams et al.，2006)。城镇化之后，不同物种经历的环境随机性增加或降低，可能会导致其种群随时间的推移而延续或局部灭绝。

2.5 总结

显然，城镇化可以破坏现存栖息地，或者改变其他栖息地，并创造新的栖息地，而且与原始地或乡村环境中没有相似之处。但是，在世界范围内与城镇化相关的初级生物物理过程和二级生物物理过程通常是相似的，而与当地植被、气候和地貌无关。微生物、真菌、植物和动物是如何响应这些过程的呢？这些过程是如何影响种群、物种和群落的呢？在城市中，生态系统维持到何种程度？在城市环境中，我们如何保护当地生物多样性？城市生活如何影响人类健康和幸福？人类和其他生物的需求在什么地方冲突或重叠？我们如何规划和建设未来的城市，为人类和其他生物提供栖息地？我们将在以下各章中解决这些问题。

📝 思考题

1. 什么是生态足迹？什么影响了一个城市的生态足迹？
2. 讨论城镇化的初级生物物理过程。你认为哪种过程的城市环境影响程度最大？
3. 什么是城镇化重要的二级生物物理过程？在城市中，这些过程是如何影响人类和其他生物的？
4. 解释城市热岛效应，以及它对城市最低、最高气温的影响。
5. 什么是环境随机性？在城市中，环境随机性是高还是低？

本章参考文献

Alig R J, Kline J D, Lichtenstein M. (2004) Urbanization on the US landscape: looking ahead in the 21st century. *Landscape and Urban Planning*, 69, 219-234.

American Forests. (2003) *Urban Ecosystem Analysis: Mecklenburg County, North Carolina*. American Forests, Washington DC. [2015-08-17]. http://charmeck.org/city/charlotte/epm/services/landdevelopment/trees/treecommission/documents/charlotte_final.pdf.

Angelakis A N, Kavoulaki E, Dialynas E G. (2014) Sanitation and wastewater technologies in Minoan Era. In *Evolution of Sanitation and Wastewater Technologies Through the Centuries*. Angelakis A N, Rose J B eds. IWA Publishing, London, 1-24.

Angold P G, Sadler J P, Hill M O, et al. (2006) Biodiversity in urban habitat patches. *Science of the Total Environment*, 360, 196-204.

Arizona Department of Economic Security. (2006) *Arizona County Population Projections*. State Data Centre, Phoenix, Arizona, USA.

Bailey H, Senior B, Simmons D, et al. (2010) Assessing underwater noise levels during pile-driving at an offshore windfarm and its potential effects on marine mammals. *Marine Pollution Bulletin*, 60, 888-897.

Baker L A, Brazel A J, Selover N, et al. (2002) Urbanization and warming of Phoenix

(Arizona, USA): Impacts, feedbacks and mitigation. *Urban Ecosystems*, 6, 183-203.

Balling R C, Gober P, Jones N. (2008) Sensitivity of residential water consumption to variations in climate: An intraurban analysis of Phoenix, Arizona. *Water Resources Research*, 44, 1-11.

BBC News, June 14, 2007; [2015-02-11]. http://news.bbc.co.uk/2/hi/africa/6754127.stm.

Beckmann J, Berger J. (2003) Rapid ecological and behavioural changes in carnivores: the responses of black bears (*Ursus americanus*) to altered food. *Journal of Zoology*, 261, 207-212.

Bellwood P. (2004) *The First Farmers: The Origins of Agricultural Societies*. Blackwell Publishing, Malden.

Beyene A, Legesse W, Triest L, et al. (2009) Urban impact on ecological integrity of nearby rivers in developing countries: the Borkena River in highland Ethiopia. *Environmental Monitoring and Assessment*, 153, 461-476.

Blomqvist L, Brook B W, Ellis E C, et al. (2013) Does the shoe fit? Real versus imagined ecological footprints. *PLoS Biology*, 11(11), e1001700. doi: 10.1371/journal.pbio.1001700.

Brazel A, Gober P, Lee S, et al. (2007) Determinants of changes in the regional urban heat island in metropolitan Phoenix (Arizona, USA) between 1990 and 2004. *Climate Research*, 33, 171-182.

Brazel A, Selover N, Vose R, et al. (2000) The tale of two climates-Baltimore and Phoenix urban LTER sites. *Climate Research*, 15, 123-135.

Breuste J H, Niemelä J, Snep R P H. (2008) Applying landscape ecological principles in urban environments. *Landscape Ecology*, 23, 1139-1142.

Bridgman H, Warner R, Dodson J. (1995) *Urban Biophysical Environments*. Oxford University Press, Melbourne.

Brook R D. (2008) Cardiovascular effects of air pollution. *Clinical Science*, 115, 175-187.

Bury B. (2008) Do urban areas favour introduced turtles in western North America? In *Urban Herpetology*. Mitchell J C, Brown R E J, Bartholomew B eds. Society for the Study of Amphibians and Reptiles, Salt Lake City: 343-345.

Changnon S A. (2001) Assessment of historical thunderstorm data for urban effects: The Chicago case. *Climatic Change*, 49, 161-169.

Chin A. (2006) Urban transformation of river landscapes in a global context. *Geomorphology*, 79, 460-487.

Chow W T L, Roth M. (2006) Temporal dynamics of the urban heat island of Singapore. *International Journal of Climatology*, 26, 2243-2260.

Cilliers S S, Williams N S G, Barnard F J. (2008) Patterns of exotic plant invasions in fragmented urban and rural grasslands across continents. *Landscape Ecology*, 23, 1243-1256.

Collins J P, Kinzig A, Grimm N B, et al. (2000) A new urban ecology: Modeling human communities as integral parts of ecosystems poses special problems for the development and testing of ecological theory. *American Scientist*, 88, 416-425.

Contesse P, Hegglin D, Gloor S, et al. (2004) The diet of urban foxes (*Vulpes vulpes*) and the availability of anthropogenic food in the city of Zurich, Switzerland. *Mammalian Biology*, 69, 81-95.

Cox W. (2008) *Demographia World Urban Areas (World Agglomerations)*. Wendel Cox Consultancy, Illinois.

Dovey K, Pafka E. (2014) The urban density assemblage: Modelling multiple measures. *Urban Design International*, 19, 66-76.

Dubai Electricity and Water Authority. (2014) DEWA Sustainability Report 2014. DEWA, Dubai. [2015-07-05]. http://www.dewa.gov.ae/images/DEWA_Sustainability_report_2014.pdf.

Dubai Electricity and Water Authority. (2015a) Water statistics 2014. [2015-07-05]. https://www.dewa.gov.ae/aboutus/waterStats2014.aspx.

Dubai Electricity and Water Authority. (2015b) Electricity statistics 2014. [2015-07-05]. http://www.dewa.gov.ae/aboutus/electStats2014.aspx.

Dubai Statistics Center. (2007) Statistical Year Book 2007. Dubai Statistics Center, Dubai. [2015-06-07]. https://www.dsc.gov.ae/en-us/Publications/Pages/publication-details.aspx? PublicationId=5 &year=2007.

Dubai Statistics Centre. (2012) Statistical Yearbook 2012. Dubai Statistics Center, Dubai. [2015-06-07]. https://www.dsc.gov.ae/en-us/Publications/Pages/publication-details.aspx? PublicationId=5&year=2012.

Dubai Statistics Center. (2015) Dubai Population Clock. [2015-06-07]. https://www.dsc.gov.ae/en-us/Pages/default.aspx.

Dyson T. (2003) HIV/AIDS and urbanization. *Population and Development Review*, 29, 427-442.

Elvidge C D, Tuttle B T, Sutton P C, et al. (2007) Global distribution and density of constructed impervious surfaces. *Sensors*, 7, 1962-1979.

Er K B H, Innes J L, Martin K, et al. (2005) Forest loss with urbanization predicts bird extirpations in Vancouver. *Biological Conservation*, 126, 410-419.

Fenn M E, Baron J S, Allen E B, et al. (2003) Ecological effects of nitrogen deposition in the western United States. *BioScience*, 53, 404-420.

Fernández-Juricic E. (2000) Local and regional effects of pedestrians on forest birds in a fragmented landscape. *The Condor*, 102, 247-255.

Fernández-Juricic E, Jokimäki J. (2001) A habitat island approach to conserving birds in urban landscapes: case studies from southern and northern Europe. *Biodiversity and Conservation*, 10, 2023-2043.

Forman R T T. (2008) *Urban Regions. Ecology and Planning Beyond the City*. Cambridge University Press, Cambridge.

Forman R T T. (2014) *Urban Ecology: Science of Cities*. Cambridge University Press, Cambridge/New York.

Fuller R A, Warren P H, Gaston K J. (2007) Daytime noise predicts nocturnal singing in urban robins. *Biology Letters*, 3, 368-370.

Gago E J, Roldan J, Pacheco-Torres R, et al. (2013) The city and urban heat islands: a review of strategies to mitigate adverse effects. *Renewable Sustainable*

Energy Review, 25, 749-758.

Gammage G, Stigler M, Clark-Johnson S, et al. (2011) *Watering the Sun Corridor: Managing Choices in Arizona's Megapolitan Area*. Morrison Institute for Public Policy, Arizona StateUniversity, Tempe. [2015-07-05]. https://morrisoninstitute.asu.edu/sites/default/files/content/products/SustPhx_WaterSunCorr.pdf.

Gartland L. (2008) *Heat Islands: Understanding and Mitigating Heat in Urban Areas*. Earthscan Publications, London.

Gaston K J, Bennie J, Davies T W, et al. (2013) The ecological impacts of nighttime light pollution: a mechanistic appraisal. *Biological Reviews*, 88, 912-927.

Gibb H, Hochuli D F. (2002) Habitat fragmentation in an urban environment: large and small fragments support different arthropod assemblages. *Biological Conservation*, 106, 91-100.

Graham A L, Cooke S J. (2008) The effects of noise disturbance from various recreational boating activities common to inland waters on the cardiac physiology of a freshwater fish, the largemouth bass (*Micropterus salmoides*). *Aquatic Conservation: Marine and Freshwater Ecosystems*, 18, 1315-1324.

Grimm N B, Faeth S H, Golubiewski N E, et al. (2008a) Global change and the ecology of cities. *Science*, 319, 756-760.

Grimm N B, Foster D, Groffman P, et al. (2008b) The changing landscape: ecosystem responses to urbanization and pollution across climatic and societal gradients. *Frontiers in Ecology and the Environment*, 6, 264-272.

Guhathakurta S, Gober P. (2007) The impact of the Phoenix urban heat island on residential water use. *Journal of the American Planning Association*, 73, 317-329.

Hamberg L, Lehvavirta S, Malmivaara-Lamsa M, et al. (2008) The effects of habitat edges and trampling on the understorey vegetation in urban forests in Helsinki, Finland. *Applied Vegetation Science*, 11, 83-98.

Harada H, Dong N T, Matsui S. (2008) A measure of provisional and urgent sanitary improvement in developing countries: septic tank performance improvement. *Water Science and Technology*, 58, 1305-1311.

Harlan S L, Brazel A J, Jenerette G D, et al. (2007) In the shade of affluence: the inequitable distribution of the urban heat island. *Research in Social Problems and Public Policy*, 15, 173-202.

Haviland-Howell G, Frankel A S, Powell C M, et al. (2007) Recreational boating traffic: a chronic source of anthropogenic noise in the Wilmington, North Carolina intracoastal water-way. *Journal of the Acoustical Society of America*, 122, 151-160.

Henry S, Boyle P, Lambin E F. (2003) Modelling inter-provincial migration in Burkina Faso, West Africa: the role of sociodemographic and environmental factors. *Applied Geography*, 23, 115-136.

Hettige H, Mani M, Wheeler D. (1998) *Industrial Pollution in Economic Development: Kuznets Revisited*. The World Bank, Development Research Group, Washington DC.

Hoornweg D, Bhada-Tata P. (2012) *What a Waste—A Global Review of Solid Waste Management*. World Bank, Washington DC.

Howard L. (1833) *Climate of London Deduced From Meteorological Observations*, vol. 1. Harvey and Darton, London.

Hynes L N, McDonnell M J, Williams N S G. (2004) Measuring the success of urban riparian revegetation projects using remnant vegetation as a reference community. *Ecological Management and Restoration*, 5, 205-209.

Ishan A I, Norddin N A M, Malek N A, et al. (2014) The quality of housing environment and green open space towards quality of life. In *Fostering Ecosphere in the Built Environment*, *UMRAN*2014. Bakar A A, Malek N A eds. International Islamic University Malaysia, Kuala Lumpur: 183-198.

Kentula M E, Gwin S E, Pierson S M. (2004) Tracking changes in wetlands with urbanization: sixteen years of experience in Portland, Oregon, USA. *Wetlands*, 24, 734-743.

Khan S. (2014) Sanitation and wastewater technologies in Harappa/Indus Valley Civilization (ca. 2600-1900 BC). In *Evolution of Sanitation and Wastewater Technologies Through the Centuries*. Angelakis A N, Rose J B eds. IWA Publishing, London: 25-42.

Krivoshto I N, Richards J R, Albertson T E, et al. (2008) The toxicity of diesel exhaust: implications for primary care. *Journal of the American Board of Family Medicine*, 21, 55-62.

Kromroy K, Ward K, Castillo P, et al. (2007) Relationships between urbanization and the oak resource of the Minneapolis/St. Paul metropolitan area from 1991 to 1998. *Landscape and Urban Planning*, 80, 375-385.

Lacoste E, Chalmin P. (2007) *From Waste to Resource*: 2006 *World Waste Survey*. Economica, Paris.

Lande R. (1993) Risks of population extinction from demographic and environmental stochasticity and random catastrophes. *The America Naturalist*, 142, 911-927.

Landsberg H E. (1981) *The Urban Climate*. Academic Press, New York.

Leary E, McDonnell M. (2001) Quantifying public open space in metropolitan Melbourne. *Australian Parks and Leisure*, 4, 34-36.

Lesage V, Barrette C, Kingsley M C S, et al. (1999) The effects of vessel noise on the vocal behaviour of belugas in the St. Lawrence River Estuary. *Canada Marine Mammal Science*, 15, 65-84.

Liu J, Zhan J, Deng X. (2005) Spatio-temporal patterns and driving forces of urban land expansion in China during the economic reform era. *Ambio*, 34, 450-455.

Löfvenhaft K, Björn C, Ihse M. (2002) Biotope patterns in urban areas: a conceptual model integrating biodiversity issues in spatial planning. *Landscape and Urban Planning*, 58, 223-240.

Löfvenhaft K, Runborg S, Sjögren-Gulve P. (2004) Biotope patterns and amphibian distribution as assessment tools in urban landscape planning. *Landscape and Urban Planning*, 68, 403-427.

Longcore T, Rich C. (2004) Ecological light pollution. *Frontiers in Ecology and the Environment*, 2, 191-198.

Luck G W, Ricketts T H, Daily G C, et al. (2004) Alleviating spatial conflict between people and biodiversity. *Proceedings of the National Academy of Sciences of the United States of America*, 101, 182-186.

Luck M, Wu J. (2002) A gradient analysis of urban landscape pattern: a case study from the Phoenix metropolitan region, Arizona, USA. *Landscape Ecology*, 17, 327-339.

Lyon L, Driskell R. (2011) *The Community in Urban Society*. Waveland Press, Long Grove.

Macionis J J, Parrillo V N. (2012) *Cities and Urban Life*. 6th edn. Pearson, Upper Saddle River.

Matsuoka R H, Kaplan R. (2008) People needs in the urban landscape: Analysis of Landscape and Urban Planning contributions. *Landscape and Urban Planning*, 84, 7-19.

May R. (1974) *Stability and Complexity in Model Ecosystems*. Princeton University Press, Princeton.

McDonald R I, Forman R T T, Kareiva P M. (2010) Open space loss and land inequality in United States' Cities, 1990-2000. *PLoS One*, 5(3): e9509. doi: 10.1371/journal.pone.0009509.

McGranahan G, Balk D, Anderson B. (2007) The rising tide: assessing the risk of climate change and human settlements in low elevation coastal zones. *Environment and Urbanization*, 19, 17-37.

McGranahan G, Marcotullio P, Bai X, et al. (2005) Urban systems, in Ecosystems and Human Wellbeing: *Current Status and Trends*. Hassan R, Scholes R, Ash N eds. Island Press, Washington DC, 795-825.

McKinney M L. (2002) Urbanization, biodiversity, and conservation. *BioScience*, 52, 883-890.

McMichael A J. (2000) The urban environment and health in a world of increasing globalization: issues for developing countries. *Bulletin of the World Health Organization*, 78, 1117-1127.

Meredith W. (2005) Mosquito control: balancing public health and the environment in Delaware. *NOAA Coastal Services Magazine*: September/October 2005. [2015-07-18]. http://www.csc.noaa.gov/magazine/2005/05/article2.html.

Morris D W, Kingston S R. (2002) Predicting future threats to biodiversity from habitat selection by humans. *Evolutionary Ecology Research*, 4, 787-810.

Murgui E. (2007) Effects of seasonality on the species-area relationship: A case study with birds in urban parks. *Global Ecology and Biogeography*, 16, 319-329.

Nassauer J I. (2004) Monitoring the success of metropolitan wetland restorations: cultural sustainability and ecological function. *Wetlands*, 24, 756-765.

Niemela J. (1999) Ecology and urban planning. *Biodiversity and Conservation*, 8, 119-131.

Njeru J. (2006) The urban political ecology of plastic bag waste problem in Nairobi, Kenya. *Geoforum*, 37, 1046-1058.

Nowak J N, Rowntree R A, McPherson E G, et al. (1996) Measuring and analyzing urban tree cover. *Landscape and Urban Planning*, 36, 49-57.

Oke T R. (1982) The energetic basis of the urban heat island. *Quarterly Journal of the Royal Meteorological Society*, 108, 1-24.

Ouyang Z, Xiao R B, Schienke E W, et al. (2008) Beijing urban spatial distribution and resulting impacts on heat islands. In *Landscape Ecological Applications in Man-influenced Areas: Linking Man and Nature Systems*. Hong S K, Nakagoshi N, Fu B J, Morimoto Y eds. Springer, Dordrecht: 459-478.

Pacione M. (2005) City profile Dubai. *Cities*, 22, 255-265.

Pargal S, Hettige H, Singh M, et al. (1997) Formal and informal regulation of industrial pollution: comparative evidence from Indonesia and the United States. *The World Bank Economic Review*, 11, 433-450.

Parés-Ramos I K, Gould W A, Aide T M. (2008) Agricultural abandonment, suburban growth, and forest expansion in Puerto Rico between 1991 and 2000. *Ecology and Society*, 13(2), 1. http://www.ecologyandsociety.org/vol13/iss2/art1/.

Parris K M, Velik-Lord M, North J M A. (2009) Frogs call at a higher pitch in traffic noise. *Ecology and Society*, 14(1), 25. http://www.ecologyandsociety.org/vol14/iss1/art25/.

Patricelli G L, Blickley J L. (2006) Avian communication in urban noise: causes and con-sequences of vocal adjustment. *The Auk*, 123, 639-649.

Pauchard A, Aguayo M, Pena E, et al. (2006) Multiple effects of urbanization on the biodiversity of developing countries: The case of a fast-growing metropolitan area (Concepcion, Chile). *Biological Conservation*, 127, 272-281.

Pearce F. (1996) How big can cities get? *New Scientist*, 190, 10.

Rebele F. (1994) Urban ecology and special features of urban ecosystems. *Global Ecology and Biogeography Letters*, 4, 173-187.

Relyea R A. (2005) The lethal impacts of roundup on aquatic and terrestrial amphibians. *Ecological Applications*, 15, 1118-1124.

Riley S P D, Busteed G T, Kats L B, et al. (2005) Effect of urbanization on the distribution and abundance of amphibian and invasive species in southern California streams. *Conservation Biology*, 19, 1894-1907.

Rosenzweig M L. (2003) *Win-Win Ecology: How the Earth's Species Can Survive in the Midst of Human Enterprise*. Oxford University Press, Oxford.

Sattler T, Borcard D, Arlettaz R, et al. (2010) Spider, bee, and bird communities in cities are shaped by environmental control and high stochasticity. *Ecology*, 91, 3343-3353.

Scalenghe R, Marsan F A. (2008) The anthropogenic sealing of soils in urban areas. *Landscape and Urban Planning*, 90, 1-10.

Seto K C, Güneralp B, Hutyra L R. (2012) Global forecasts of urban expansion to 2030 and direct impacts on biodiversity and carbon pools. *Proceedings of the National Academy of Sciences*, 40, 16083-16088.

Shochat E. (2004) Credit or debit? Resource input changes population dynamics of city-slicker birds. *Oikos*, 106, 622-626.

Shrestha M K, York A M, Boone C G, et al. (2012) Land fragmentation due to rapid urbanization in the Phoenix Metropolitan Area: analysing the spatiotemporal patterns and drivers. *Applied Geography*, 32, 522-531.

Shukuroglou P, McCarthy M A. (2006) Modelling the occurrence of rainbow lorikeets

（Trichoglossus haematodus）in Melbourne. *Austral Ecology*, 31, 240-253.

Slabbekoorn H, Yeh P, Hunt K. (2007) Sound transmission and song divergence: a comparison of urban and forest acoustics. *The Condor*, 109, 67-78.

Spinks P Q, Pauly G B, Crayon J J, et al. (2003) Survival of the western pond turtle (*Emys marmorata*) in an urban California environment. *Biological Conservation*, 113, 257-267.

Sturman A P, Tapper N J. (2006) *Weather and Climate in Australia and New Zealand*. Oxford University Press, Oxford.

Suren A M, Riis T, Biggs B J F, et al. (2005) Assessing the effectiveness of enhancement activities in urban streams: I. Habitat responses. *River Research and Applications*, 21, 381-401.

Tait C, Daniels C B, Hill R S. (2005) The urban ark: The historical evolution of the plant community. In *Adelaide. Nature of a City: The Ecology of a Dynamic City from 1836 to 2036*. Daniels C B, Tait C eds. BioCity, Adelaide, 87-110.

Tan J, Zheng Y, Tang X, et al. (2010) The urban heat island and its impacts on heat waves and human health in Shanghai. *International Journal of Biometeorology*, 54, 75-84.

Torok S J, Morris C J G, Skinner C, et al. (2001) Urban heat island features of southeast Australian towns. *Australian Meteorological Magazine*, 50, 1-13.

UAE Interact. (2007) *UAE Yearbook* 2007. [2014-02-24]. http://www.uaeyearbook.com/Yearbooks/2007/ENG/.

UNEP. (2005) *Selection, Design and Implementation of Economic Instruments in the Solid Waste Management Sector in Kenya: The Case of Plastic Bags*. UNEP *Division* of Technology, Industryand Economics, Paris. [2015-07-18]. http://www.unep.ch/etb/publications/econinst/kenya.pdf.

UNFPA. (2007) *State of World Population* 2007. *Unleashing the Potential of Urban Growth*. United Nations Population Fund, New York.

US Energy Information Administration. (2015) Arizona State Profile and Energy Estimates. [2015-06-07]. http://www.eia.gov/state/?sid=AZ.

US Environmental Protection Agency. (2014) *Municipal Solid Waste Generation, Recycling, and Disposal in the United States: Facts and Figures for* 2012. [2015-06-07]. http://www.epa.gov/osw/nonhaz/municipal/pubs/2012_msw_fs.pdf.

Van de Poel E O'Donnell O, Van Doorslaer E. (2009) What explains the rural-urban gap in infant mortality: household or community characteristics? *Demography*, 46, 827-850.

Walsh C J, Papas P J, Crowther D, et al. (2004) Storm water drainage pipes as a threat to a stream-dwelling amphipod of conservation significance. *Austrogammarus australis*, in south-eastern Australia. *Biodiversity and Conservation*, 13, 781-793.

Walsh C J, Roy A H, Feminella J W, et al. (2005) The urban stream syndrome: current knowledge and the search for a cure. *Journal of the North American Benthological Society*, 24, 706-723.

Warren P S, Katti M, Ermann M, et al. (2006) Urban bioacoustics: it's not just noise. *Animal Behaviour*, 71, 491-502.

WHO. (2014a) Ambient air pollution database, May 2014. [2015-06-08]. http://www.who.int/phe/health_topics/outdoorair/databases/cities/en/.

WHO. (2014b) Ambient(outdoor) air quality and health. Fact Sheet No. 313, updated March 2014. [2015-07-18]. http://www.who.int/mediacentre/factsheets/fs313/en/.

WHO, UN-Habitat. (2010) *Hidden Cities: Unmasking and Overcoming Health Inequities in Urban Settings*. WHO Press, Geneva.

Williams N S G, McDonnell M J, Phelan G K, et al. (2006) Range expansion due to urban-ization: Increased food resources attract Grey-headed Flying-foxes (*Pteropus poliocephalus*) to Melbourne. *Austral Ecology*, 31, 190-198.

WMO. (2015a) World Weather information Service: Dubai, United Arab Emirates. http://worldweather.wmo.int/en/city.html?cityId=1190 (accessed on 7/06/2015).

WMO. (2015b) World Weather information Service: Phoenix, Arizona. http://worldweatherwmo.int/en/city.html?cityId=806 (accessed on 7/06/2015).

World Bank. (2013) *Global Monitoring Report* 2013: *Rural-Urban Dynamics and the Millennium Development Goals*. World Bank, Washington DC.

WWF. (2014) *Living Planet Report* 2014: *Species and Spaces*. People and Places, WWF, Gland, Switzerland.

Xiao R, Ouyang Z, Zheng H, et al. (2007) Spatial pattern of impervious surfaces and their impacts on land surface temperature in Beijing, China. *Journal of Environmental Sciences*, 19, 250-256.

第3章 种群和物种水平对城镇化的响应

3.1 概述

城镇化的生物物理过程是通过改变微生物、真菌、植物和动物等赖以生存的资源的数量、质量、时间和空间分布而对种群和物种产生广泛影响，例如：庇护、筑巢、食物、水、阳光和营养（Riley et al.，2003；Shochat et al.，2006；Harper et al.，2008；McDonald et al.，2008）。城市还经历了许多生物引入和入侵，这些物种散布到城市中是人类有意或无意造成的（Rebele，1994；Lambdon et al.，2008；Williams et al.，2009）。另外，城市区域人群高密度，增加了干扰其他生物以及其栖息地的可能性。常见的干扰包括践踏植物和动物，采集真菌和植物，猎食动物以及改变对人类敏感的动物行为方式（Fernández-Juricic，2000；Milner-Gulland et al.，2003；Murison et al.，2007；Peres，Palacious，2007；Florgård，2009）。这些过程（城镇化的初级、二级生物物理过程、生物引进和人类干扰），共同影响了城市环境中其他生物的多度，改变了物种内部和物种之间的相互作用，包括争夺资源、捕食、自相残杀、同化，以及更复杂的相互作用，例如：捕食者介导的竞争。

种群增长或下降与四个过程或存活率相关：出生、死亡、迁入和迁出。因此，研究种群（以及所属物种）如何响应城镇化的逻辑方法是应该考虑城镇化如何影响种群动态率，最终导致种群是增长还是下降。图3.1显示了概念模型的重要机制和途径，与城镇化相关的环境变化可以通过这些机制和途径影响生物种群。人为干扰、资源利用率的变化以及种内、种间相互作用的变化，影响幼年和成年生物的生存、繁衍后代以及分散在城市中或远离城市地区的种群数量。环境

3.1 概述

和统计的随机性、灾害性事件（如第2章所定义）也可能影响城市环境中生物的存活率。随后种群的增长或下降可以进一步改变物种间的相互作用，建立一个反馈循环，不同物种的种群通过该反馈循环继续增加或减少，有时甚至达到局部灭绝的地步（图3.1）。例如，由于加利福尼亚州旧金山市的城市发展，泽西斯蓝小灰蝶（*Glaucopsyche xerces*）被认为是第一种灭绝的美国蝴蝶（Pyle，1995）。

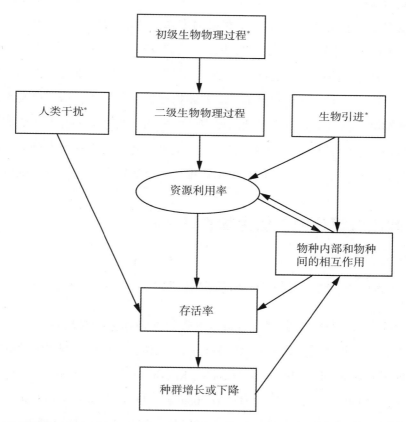

图 3.1　与城镇化过程相关的概念模型，以及这些过程可以影响物种种群的重要机制和途径
（＊表示受人类偏好强烈影响的过程）

图 3.1 中用星号标出与城市建设和扩张有关的三组过程，这些过程受到人类偏好的强烈影响，即城镇化进程中，初级生物物理过程、人类干扰和生物引进。审美、文化偏好、人口压力、社会经济因素和城市发展速度会影响城市的自然形态，例如：清除了多少植被，有多少开敞空间，有多少可渗透表面，道路和房子

51

密度，自然地物如湿地是否保留、改变或破坏等（Sharpe et al.，1986；Montgomery，1998；Clark et al.，2002；Kentula et al.，2004）。人类的偏好也会影响城市不同地区的活动模式，如参观哪个地方，何时、有多少人以及他们在那做什么（Burgess et al.，1988；Storper，Manville，2006）。

　　大自然爱好者最有可能在春季和夏季走过公园或原始植被的残余斑块，从而有可能践踏植物或者干扰正在繁殖的鸟类。城市居民可能会捕采野生动植物作为食物，或在市场上出售，或采集立木和倒木作为燃料（Maharjan，1998；Asfaw，Tadesse，2001）。此外，人类的偏好决定了人们有意引入城市生活的外来物种类型（Hope et al.，2003；Muerk et al.，2009）。各种各样的栽培园艺植物，包括乔木、灌木、花卉、草本植物和蔬菜，家养动物如猫、狗、鸡、猪，以及观赏鱼类如金鱼和锦鲤，已被引入世界各地的城市。家猫和狗几乎是无所不在的城市居住者；来自南非的普罗蒂斯（Proteas）花，日本的杜鹃花，可以在澳大利亚、欧洲和北美洲的城市庭院中见到。下面将更加详细地研究生物种群如何响应城镇化。

3.2　响应城镇化的二级生物物理过程

3.2.1　栖息地消失、破碎化和孤岛化

　　城镇化之后，许多微生物、真菌、植物和动物在其栖息地消失（生境丧失）后会死亡，尽管某些微生物和真菌物种可能会残留在土壤中，某些类型的动物个体也可能成功地散布到附近地区（Czech，Krausman，1997；Czech et al.，2000；How，Dell，2000；Faulkner，2004；Newbound，2008；Newbound et al.，2010）。因此，生境丧失对种群的影响是显著而直接的。相比之下，栖息地破碎化（将剩余栖息地分割成小片）和栖息地隔离的影响可能更加微妙，需要种群多代繁衍后才能体现出来（表3.1）。集合群落理论表明（Levins，1969），被改变的景观所包围的栖息地斑块可能类似于海洋岛屿。该理论预测，小的斑块相对于大斑块来说，不太可能支撑一个特定的物种的种群，因为局部灭绝的可能性较高；而孤立的斑块比相邻其他适宜的斑块更不可能支撑一个种群，因为移居到另一个斑块的可能性较低（Hanski，1994，1998；图3.2）。因此，随着时间的流逝，栖居在被

城市包围的狭小、孤立栖息地中的种群可能会继续消失，这个观点被称为"灭绝债务"（extinction debt）（Tilman et al.，1994；McCarthy et al.，1997）。由于景观变化和种群响应之间存在时间差，因此，世界上仍然存在大量的灭绝债务，是过去全世界城市中自然栖息地的破碎化和孤岛化所付出的代价（Hanski，Ovaskainen，2002；Hahs et al.，2009；Kuussaari et al.，2009）。因此，目前栖息地破碎斑块中的种群存在并不能很好地表明它们在未来会持续存在，特别是长寿命物种（Spinks et al.，2003）。

表 3.1 在城市环境中种群和物种的栖息地消失、破碎化和孤岛化的主要影响

栖息地消失、破碎化和孤岛化对种群和物种的影响
栖息地消失会导致：
个体存活率下降
个体迁出
种群局部灭绝
栖息地破碎化会导致：
较小的栖息地，支持较小的种群
局部灭绝的可能性更高
降低了物种的丰富度
栖息地孤岛化会导致：
减少个体在栖息地之间移动
随着局部灭亡，定居或重新定居的可能性较低
授粉率较低，破坏种子和孢子的扩散
栖息地之间基因流减少，遗传多样性丧失

为了保持这种种群（或破碎化）模型，栖息地的残余斑块需要被敌对的非栖息地所包围，它不提供食物或庇护这样的资源，并且穿越非栖息地困难或代价高（Bun-nell，1999；Lindenmayer，Franklin，2002）。对于需要特定资源或条件的物种，这些条件可能会满足，而这些条件在指定栖息地之外的城市地区并不存在（参见 Heard et al.，2013），或满足无法在生境斑块之间成功扩散的物种。栖息地被非常密集的城市地区（不透水表面的覆盖率高）所包围（几乎没有为任何物种

提供资源的地区），也更容易出现这种情况。Drinnan（2005）发现，悉尼南部的大型林地保护区比小型保护区支持更多种类的真菌、植物、鸟类和青蛙，而相比那些有栖息地廊道连接或紧邻其他大的保护区，更多孤立保护区支持的物种更少。在对英格兰萨塞克斯城市两栖动物的早期研究中，随着城市发展密度的增加，青蛙、蟾蜍和蝾螈几乎不可能出现在花园池塘中（Beebee，1979）。

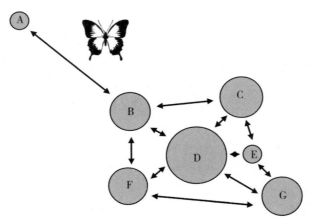

图3.2 假设蝴蝶栖息地网络由7个大小不同且彼此接近的栖息地斑块组成（圆形）；箭头显示了在斑块之间扩散的途径。种群理论预测，相比大斑块和相连接的斑块，小斑块、孤立斑块支撑的种群较小。在这个案例中，斑块A（小和孤立）是最不可能支持一个种群的，而斑块D（斑块大和较好的连通性）则最有可能支持一个种群。

连续模型可以用来替代城市区域的栖息地和非栖息地的二元分类，该模型允许空间中不同物种所需资源逐渐变化（Fischer，Lindenmayer，2006）。对于栖居在城市和非城市环境中的一系列物种而言，这可能是一个更适合的模型。例如，一些原生的负鼠物种在澳大利亚东部的城市中很丰富，并且在城市中可达到比非城市生境中更高的密度（Harper，2005）。一项对墨尔本残余林地中常见的浣熊和帚尾负鼠的研究发现，它们的多度随着斑块内和市区周围资源的可利用率的增加而增加（Harper et al.，2008）。在林地斑块中，每种负鼠种群的密度较高，这些林地中多有潜在的巢穴点（空心树），并且在巢穴点周围100m的缓冲区中乔木和灌木的覆盖率高。林地斑块外主要由外来园林植物组成，相比已经进化出化学防御的本地植物，这些植物对负鼠来说更美味和更有营养。

在缺少空心树的林地斑块中，缓冲区中可食树种的覆盖率对负鼠的多度影响不大，看来负鼠需要足够的巢穴，以显示种群水平响应城市环境中食物可利用性的增加。一个资源控制实验中，短短 6 个月内，研究人员用苹果补充喂养帚尾负鼠，在 20 个林地斑块中帚尾负鼠的平均数量增加了 1 倍（Harper，2005）。考虑到种群的响应速度和繁殖季节的时间选择，从周围区域迁入，而不是存活率或繁殖力的增加，最有可能是补充食物使负鼠多度增加。在这个例子中，负鼠可以轻易越过特定的栖息地斑块边界获取临近区域的资源，且城市环境相对良性（郊区开发中建有绿树成荫的花园）。

当栖息地被分割成较小的区域时，它们会随着时间的流逝进一步发生变化。小斑块栖息地易受边缘效应的影响，在两种生态系统或栖息地类型的边界发生物理和生物条件的变化。在陆地生境中，这些条件的变化包括微气候的变化（如辐射、温度和风力的增加，湿度降低等），外来物种入侵，来自食草动物和食肉动物的威胁增加，污染物的流入和人类干扰的增加（Morgan，1998；Beale，Monaghan，2004；Zhou et al.，2004；Grimm et al.，2008）。城市环境中受干扰或破碎化的水生生境也遭受边缘效应的影响，例如：外来物种的入侵、捕食压力增加以及水文和热环境改变（Temple，1987；King，Buckney，2000；Trombulak，Frissell，2000；Konrad，Booth，2005）。多数变化因素都会改变不同物种之间的资源可用性，物种之间的互动，或直接影响小斑块的生命率。在光照充足的条件下，栖息地斑块的边缘可用资源可以增加某些植物的竞争优势（Hamberg et al.，2008）。相反，栖息地边缘存在较高密度的鸟类捕食者，会增加其他鸟类的捕食压力，从而降低繁殖率（Gates，Gysel，1978；Batary，Baldi，2004）。人类干扰，包括采集薪柴、砍伐或踩踏植被，可降低小栖息地复杂性，破坏了为无脊椎动物和小脊椎动物提供庇护的结构组成，如本土的啮齿动物（Soulé et al.，1992；Sauvajot et al.，1998）。

城市区域高密度道路对许多陆地动物，包括节肢动物、两栖动物、爬行动物和哺乳动物，形成了实质性的障碍，孤立了城市栖息地的种群（Gibbs，Shriver，2002；Marchand，Litvaitis，2004；Seiler et al.，2004；Vandergast et al.，2007；Noel et al.，2007；Hale et al.，2013）。当动物必须穿越马路才能到达目的地时，它们的日常活动、季节性迁徙、远距离扩散都会受到限制。Hels 和 Buchwald

(2001)证明,试图穿越马路的青蛙和蝾螈可能会遭受很高的死亡率。尽管它们能够安全穿过安静的道路,但是随着交通量的增加,成功通过的可能性急速下降。例如,一种常见的蝾螈或有羽冠的蝾螈,穿越一条每天承载1万辆车的道路,安全通过率会低于20%(Hels, Buchwald, 2001)。当两栖动物穿越城市密集道路区时,安全穿越多条道路的可能性几乎为零。在英国,道路交通是造成狗獾死亡的最大因素(Clarke et al., 1998),繁忙的高速公路也成为哺乳动物扩散的主要障碍物,例如:美国的土狼(图3.3)和鲁弗斯山猫(Riley et al., 2006)。

图3.3 繁忙的高速公路成为一个障碍物,阻碍了野狼的扩散(https://pixabay.com/en/coyote-penn-dixie-hamburg-new-1820/.)

道路和其他城市基础设施对动物活动的干扰,将会影响一些真菌和植物依赖某些无脊椎动物或脊椎动物进行授粉、散布孢子或种子,影响这些真菌和植物种群的持续性(Bhattacharya et al., 2003;Close et al., 2006;Newbound et al., 2010;Bates et al., 2011)。道路也是甲虫、大黄蜂、蜻蜓和蝴蝶等许多昆虫安全移动的障碍(Muñoz et al., 2015)。蝴蝶经常被公路上行驶的车辆杀死;通常,被杀死的个体和物种数量随着交通量和道路宽度的增加而增加(McKenna et al., 2001;Rao, Girish, 2007; Skórka et al., 2013)。在道路上较高的死亡率也与蝴蝶的某些特点和性状有关:小体型和喜欢低飞的蝴蝶的死亡率较高(Rao, Girish, 2007; Skórka et al., 2013)。经常移动或长距离移动的蝴蝶种类可能比定居蝴蝶种类被杀死的概率更高(De la Puente et al., 2008)。在马萨诸塞州波士顿进行的一项实验发现,大黄蜂(*Bombus impatients*, *B. affinis*)可以越过道路或铁路线到达另一侧合适的觅食栖息地,但是先天的忠于固定位置的特性使得它们很少这样做,据推测由于道路的干扰,大黄蜂将这个栖息地视为一个独立的场地(Bhattacharya et al., 2003)。因此,这些人为构成物可能会限制大黄蜂的活动,并导致城镇环境中的植物群落破碎化。

城市发展导致种群孤立,体现在它们的遗传结构,也即种群之间的遗传分化。孤立种群之间减少了基因流,导致遗传结构增加,即种群之间的分化,并且在某些情况下,遗传多样性降低。在极端情况下,遗传多样性低可能导致近亲繁殖,进而降低后代的存活率和繁殖力(Reed, Frankham, 2003)。众所周知,近亲繁殖对许多野生种群产生不利影响(Hedrick, Kalinowski, 2000; Keller, Waller, 2002)。对昆虫、两栖动物和哺乳动物在内的一系列生物分类的研究表明,城市环境中的栖息地破碎化和遗传多样性减少之间存在关联(参见 Hitchings, Beebee, 1998; Robinson, Marks, 2001; McClenaghan, Truesdale, 2002; Vandergast et al., 2007)。种群小又孤立,也容易遭受遗传瓶颈的影响,这是由于有效种群数量在短期内明显下降和遗传漂移所致。在遗传漂移中,某些等位基因不会仅靠偶然机会传给下一代(Frankham et al., 2004)。遗传多样性低也会限制种群进化以应对环境变化或新型捕食者和病原体的能力(Frankham et al., 2002)。

3.2.2 气候变化

随着城镇化发展,城市的温度和湿度也在变化,会影响城市区域的真菌、植物和动物的生命周期(物候)时间。在温带气候下,最常观察到在较温暖的城市区域,植物开花、发芽和结果也较早(Roetzer et al., 2000; White et al., 2002; Menzel, 2003),春季繁殖的动物繁殖较早(参见 Luniak, 2004; Partecke et al., 2005; Chamberlain et al., 2009)。如果猎物的可供应峰值和捕食者的食物需求峰值不重合，物种对气候变暖的不同反应对营养系统产生重要影响（Durant et al.，2007；Both et al.，2008）。在城市环境中，对公园和花园进行人工灌溉可能会延长植物的开花期和结果期，减少其变异性，从而延长一年中花和果实作为其他物种的食物资源的时间。在亚利桑那州凤凰城炎热、干燥的气候中，城市热岛效应增加了夏季对乔木、灌木和冷季型草等植物的高温胁迫，同时减少了对寒冷敏感物种的冬季寒冷压力（如仙人掌、多肉植物和暖季型草）（Baker et al.，2002）。自1948 年以来，城市热岛效应还将该城市每年的节肢动物的活动期，延长了大约 1个月。

城市气候变化可能会吸引移居者来到城市，例如：澳大利亚墨尔本的灰头狐蝠（Parris，Hazell，2005）。传统上认为，该物种是温带至亚热带的物种，少数的灰头狐蝠于 1986 年在墨尔本建立了永久的（全年）栖息地，到 2003 年，种群增长到大约 3 万只（van der Ree et al.，2006）。根据长期数据推断，墨尔本不属于澳大利亚其他灰头狐蝠的安营扎寨的气候范围（Parris，Hazell，2005）。然而，由于澳大利亚观测点总体变暖（Torok，Nicholls，1996），以及城市热岛效应的增强（Torok et al.，2001；Lenten，Moosa，2003），导致自 1950 年以来，墨尔本市中心的温度一直在上升。此外，公园和花园的人工灌溉，可能相当于每年 590mm的额外降雨量（Parris，Hazell，2005）。在墨尔本市，人类活动使得气温增高和有效降雨量增大，为灰头狐蝠的栖息地创造了更适宜的气候条件。相反，城市区域的气候变化会导致失去气候生态位的本地种群灭绝。瑞士巴塞尔附近 8 个本地的蜗牛种群的灭绝与大规模的城市开发有关（Baur，Baur，1993）。蜗牛卵的成功孵化率，随着暴露在 ≥22℃ 气温下而降低，这意味着因温度升高而影响了种群的存活率，最终影响城市环境中的物种持续性。

3.2.3 改变了水文环境

城市河流、池塘和湿地水文环境的变化会影响多种水生生物，包括藻类、无脊椎动物、鱼类和两栖动物（Walsh et al.，2005；Hamer，McDonnell，2008）。城市河流的流动及其导致的河流形态和水质的变化，改变了资源的可用性，如食物、庇护所、附着和繁殖场所，改变类群之间的相互作用，并最终影响存活率。河流的高流动性可以冲洗下游的个体动物，并从河床上冲刷藻类和植物。最终，城市河流倾向于支持抗干扰的物种种群，而其他物种则被排除（Walsh et al.，2005）。伯明翰的 Tame 河的水力模型比较了河流的流速模式和三种鱼类（鲢鱼、鲦鱼和斜齿鳊）的最大可持续游动速度（Booker，2003）。在大多数可用的栖息地中，河流速度在一年内有 16 次超过了这些鱼类的最大可持续游动速度（MSSS），这表明河流流速高很可能将它们排除在河流之外。

城镇化可以改变城市中池塘和湿地的平均水文周期（存水的时间周期），水的流入或流出加深现存的池塘使它们更永久；也会破坏临时湿地，为城市开发让路（Hamer, McDonnell, 2008）。较长的水文周期有助于延长鱼类和两栖类动物幼虫生命的跨度，并与鱼类和无脊椎动物捕食者共存于永久性池塘中。相比之下，较短水文周期更利于幼虫寿命短的两栖动物，这些两栖动物更适合生活在捕食者少的短暂性池塘（Hamer, McDonnell, 2008）。在美国宾夕法尼亚中部城市，随着城镇化的发展，湿地水文周期增加（Rubbo, Kiesecker, 2005）。在研究区域中，这些城市湿地往往有鱼类，支持的两栖动物幼体往往少于乡村湿地。城市湿地中很少见到树蛙（*Rana sylvatica*）、斑点蝾螈（*Ambystoma maculatum*）、杰弗逊蝾螈（*A. jeffersonianum*），除了对城市区域森林栖息地的丧失做出反应外，它们的幼体也易受其他鱼类捕食（Rubbo, Kiesecker, 2005）。在这种情况下，城镇化协同二级生物物理变化（水文周期增加）增加了鱼类和某些两栖动物物种的关键资源可用性，进而对它们的生存和繁殖产生了积极影响。随着种群的繁荣，池塘中这些物种与其他物种之间的相互作用也发生了变化。较高的鱼类密度增加了脆弱两栖动物（适应于没有鱼类的短暂池塘的两栖动物）幼虫的被捕食率，导致它们的种群下降（图 3.1）。适应永久性水体的两栖动物幼体，可能会比适应短暂性生境的两栖动物缺少食物竞争（Wellborn et al.，1996）。

3.2.4　大气、水和土壤污染

城市污染和废物对陆地、水生生境的物种和种群产生影响。在许多情况下，这些污染物和废弃物是有毒物质的来源，如重金属和有机化合物，它们会增加生物的发病率和死亡率（Snodgrass et al.，2008；Foster et al.，2014；Morrissey，2014）。但是，其他无毒废弃物也会为物种提供可以利用的资源，如营养物质、食物和庇护（Gehrt，2004；Grimm et al.，2008）。对污染比较敏感的物种种群数量往往会减少，由于较低的存活率，在污染环境中它们与其他物种竞争的能力也比较弱，而耐污染和废弃物的物种可能会繁衍生息（Agneta，Burton，1990；Walsh et al.，2007；Yule et al.，2015）。大型水生无脊椎动物对污染的反应众所周知，它们不同种群的出现和多度是水质的可靠指标（Resh，Jackson，1993；Johnson et al.，2013）。

最近对埃塞俄比亚博尔肯纳河的一项城市污染研究发现，大型无脊椎动物种群与上游对照点的无脊椎动物种群发生了显著变化（Beyene et al.，2009）。无脊椎动物的总体多度和现存物种数量随着水污染的加剧而减少，从而导致摇蚊和污泥虫占主导地位。受城市污染影响，测试点的氮、磷含量高，而溶解氧含量低。两个测试点5天的生化需氧量BOD_5（微生物在5天内分解水中有机物所需要的溶解氧量）超过1100mg/L。相比之下，为保护养殖鱼类、软体动物和甲壳类动物免受生化胁迫，推荐BOD_5的水平<15mg/L（ANZECC，2000）。埃塞俄比亚的德西埃和孔博勒查没有正式污水处理系统或污水处理厂，工业和城市废水直接排入河流；这些城市居民还将未经处理的河水用于饮用和沐浴，这对公共健康具有明显的风险（Beyene et al.，2009）。

许多附生藻类、地衣和苔藓植物对空气污染敏感，特别是对硫和氮氧化物的含量敏感（Gilbert，1969；Hawksworth，1970；Leblanc，DeSloover，1970；Puckett et al.，1973；Giordani，2007；Gadsdon et al.，2010）。在城市区域，树木上的敏感类群的消失可能对以附生植物作为食物或庇护所的无脊椎动物产生重要的次要作用。20世纪60年代，在英国纽卡斯尔进行的一项研究中，Gilbert（1971）发现，越接近市中心（SO_2浓度达到年均$224\mu g/m^3$），白蜡树上藻类和地衣的覆盖度、多样性与空气污染增加有关。随着空气污染的加剧，以这些藻类和地衣为食的树虱的多度和物种丰富度也下降了（图3.4）。然而，近几十年来，欧洲空气质量的改善可能正帮助城市地衣群落恢复（Lättman et al.，2014）。例如，芬兰坦佩雷的年平均大气SO_2浓度从1973年的$160\mu g/m^3$下降到1999年的$2\mu g/m^3$，这与椴树（*Tilia x vulgaris*）树干上附生地衣的物种丰富度和覆盖度的增加相吻合（Ranta，2001）。在1982年至2000年，25个研究地点中平均物种丰富度从0.7种增加到7.6种，而地衣的平均覆盖度从0.06%增加到10.9%。

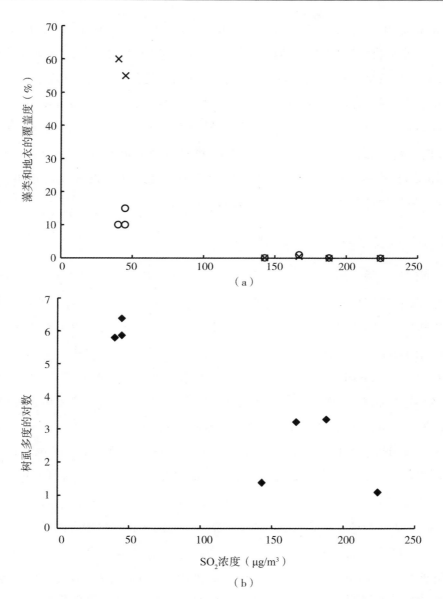

图 3.4 （a）树干上的藻类和地衣的覆盖度(%)；（b）英国纽卡斯尔的树虱多度的对数与年平均大气 SO_2 浓度的关系； ○ = 内白斑球菌(*Pleurococcus naegli*)的覆盖度(%)； × = *Lecanora Conizaeoides* 地衣的覆盖度(%)（数据来源：Gilbert，1971）

生活垃圾（城市垃圾）为许多栖居在城市中的动物提供了重要的食物资源，这些动物包括蜥蜴、老鼠、猫、狐狸、土狼、狒狒和熊（Banks et al.，2003；

Beckmann，Berger，2003；Contesse et al.，2004；Fox，2006；Powell，Henderson，2008）。与动物的天然食物资源不同的是，这些生活垃圾是一年四季都可以利用，高度集中且源源不断（Beckmann，Berger，2003）。这些"资源"增加它们的生存和繁殖力，并最终提高了利用该"资源"的物种的种群数量（Robbins et al.，2004；McKinney，2002）。有趣的是，吃垃圾的动物每天的活动时间，往往比依赖自然食物的动物少，因为它们花费更少的时间即可完成觅食（Banks et al.，2003；Beckmann，Berger，2003）。这可能会导致健康问题，如肥胖和胰岛素阻抗，肯尼亚吃垃圾的狒狒就存在这个问题（Banks et al.，2003）。

3.2.5 改变了噪声和光环境

通信是动物之间所有社会关系的基础。昆虫、鱼类、蛙类、鸟类和哺乳动物使用声音信号实现多种社会目的的交流，包括乞求食物，吸引伴侣并建立联系，捍卫领土，与团体保持联系，对靠近掠食者的危险警告。道路交通噪声或者娱乐游船、商业运输产生的噪声，会干扰、缩短信息传递的距离（Warren et al.，2006；Bee，Swanson，2007；Vasconcelos et al.，2007；参见专栏3.1），这称为声干扰或掩蔽。过往车辆的噪声还会使动物警觉，引发生理应激反应，并以各种方式改变动物的行为（Singer，1978；Sun，Narins，2005；Samuel et al.，2005；Kight，Swaddle，2011；Shannon et al.，2014；Parris，2015）。在美国怀俄明州，一项对艾草榛鸡（*Centrocercus urophasianus*）的研究发现，雄鸟会避开有汽车喇叭声的交通噪声作为其求偶地点（这些地点可呼唤、吸引雌性交配）（Blickley et al.，2012a）。在嘈杂场地求偶的雄鸟的荷尔蒙水平，高于在安静地带的雄鸟的荷尔蒙水平（Blickley et al.，2012b）。高噪声环境还会提高动物的警觉性，因为它们听不到捕食者到来的响动，被捕食物种花费更多的时间来观察捕食者，而花更少的时间觅食（Barber et al.，2010）。道路建设和交通噪声甚至可能导致动物暂时性或永久性的听力损失，包括鱼类、爬行动物和海洋哺乳动物（Brattstrom，Bondello，1983；Erbe，2002；McCauley et al.，2003；Popper，Hastings，2009；Slabbekoorn et al.，2010）。

> 专栏 3.1

船只噪声和卢西坦尼亚蟾鱼的声通信

卢西坦尼亚蟾鱼（*Halobatrachus didactylus*）产于地中海和大西洋东部的河口和近沿海地区。在繁殖季节，雄性蟾鱼会在浅水中建立领地，并在岩石下筑巢。与雄性青蛙类似，雄性蟾鱼个体会建立彼此靠近的领土，然后使用尖锐的声调信号来吸引雌性进行交配（Dos Santos et al., 2000；Vasconcelos et al., 2007）。该信号还具有地域功能，用于在雄性与雄性相遇时，将入侵者拒之门外（Vasconcelos et al., 2010）。卢西坦尼亚蟾鱼的地理分布与繁忙的航运线重合，并且人们担心船上的低频水下噪声可能会干扰它们的声学交流。一项实验室的研究表明，渡船发出的噪声大大提高了蟾鱼的听觉阈值，缩短了它们之间听到其交配和领域信号的距离，减小为雌性提供选择伴侣的空间（Amorim，Vasconcelos，2008）。船上的噪声可能掩盖了这些信号的差异，并有可能损害了该物种的配偶选择（Vasconcelos et al., 2007）。正如我们在许多种类青蛙中所观察到的那样，卢西坦尼亚蟾鱼繁殖是否成功，与更高的呼唤率和更长的召唤时间相关（Vasconcelos et al., 2012）。较高的呼唤率也是体型较大和健康状况较好的指标。

鸣鸟表现出一系列对城市噪声的行为反应，包括鸣叫频率较高以减少来自低频噪声的干扰，更大声地鸣叫和改变昼夜鸣叫模式以避开交通高峰期（Slabbekoorn，Peet，2003；Brumm，2004；Warren et al., 2006；Wood，Yezerinac，2006；Fuller et al., 2007；Parris，Schneider，2009；Potvin et al., 2011）。最近的研究结果表明：鸣鸟可以将其声音信号的频率，这是应对城市噪声的一种高可塑性的短期行为反应（Bermúdez-Cuamatzin et al., 2009；Halfwerk，Slabbekoorn，2009；Potvin，Mulder，2013）。据观察，来自澳大利亚南部的棕色树蛙（*Litoria ewingii*）和来自德国的弓翅蚱蜢（*Chorthippus biguttulus*），在交通噪声中用更高的声调鸣叫（Parris et al., 2009；Lampe et al., 2012）。鲸鱼在游经较高噪声的船只时，它们的鸣叫频率和振幅发生了变化。在弓翅蚱蜢例子中，发育可塑性而非个体行为可塑性，似乎是路边栖息地动物的鸣叫频率变化的机制（Lampe et al., 2014；图 3.5）。

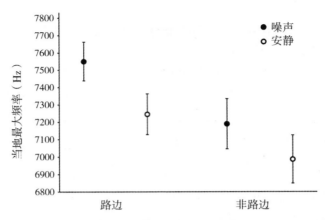

图3.5 在嘈杂环境下,弓翅蚱蜢的鸣叫频率变化(当地最大频率,单位Hz),显示了个体在路边和非路边栖息的差异,参与统计的雄性个体平均重 95.15mg (Lampe et al. , 2014,图2)

交通噪声对动物种群的影响是很难观察到的。在荷兰的一项早期研究发现,在靠近道路的栖息地中,一些鸟类的繁殖率较低,且种群密度也较低(Reijnen, Foppen,1994; Reijnen et al. ,1995,1996),尽管这些结果可能因路边研究点和对照点之间的生境差异而混淆。有人提出,交通噪声是雄柳莺在繁忙道路附近繁殖成功率降低的主要原因,因为这些雄鸟很难吸引并保持伴侣(Reijnen, Foppen, 1994)。最近对大山雀的研究发现了相似的结果:在嘈杂区域,雌性大山雀产卵的次数减少了,雏鸟的数量也减少了(Halfwerk et al. ,2011)。在澳大利亚东南部莫宁顿半岛,随着交通噪声的增加,在路边栖息地中发现灰色伯劳画眉(*Colluricincla harmonica*)和灰色扇尾水鸲(*Rhipidura fuliginosa*)的概率下降(Parris, Schneider,2009)。在58个研究点中,发现灰色伯劳画眉到访研究点的概率,从最安静站点的86%(95%置信区间:61%~100%)到最嘈杂站点的10%(0~28%)不等。考虑了研究点之间的植被类型、宽度、景观环境和交通量(通过车辆的数量)的差异时,这种关系也成立,这表明该物种出现次数减少的原因是噪声造成的。一项针对"幽灵之路"的实验发现(在美国爱达荷州的无路景观中引入交通噪声),在受到噪声影响的几天中,多种鸟类的数量较少(McClure et al. ,2013),这表明鸟类是对噪声本身,而不是对道路的任何其他特征做出的反应。

在美国新墨西哥州的林地中,某些鸟类在受到天然气压缩机噪声影响的地点筑巢的频率要比在安静的对照点低，而其他一些鸟类在此地的筑巢率则更高（Francis et al.，2009）。黑头蜡嘴雀（*Pheucticus melanocephalus*）的所有巢穴，以及哀鸽（*Zenaida macroura*）的 23 个巢穴中有 22 个都是位于安静的地方。这两种鸟的沟通方式都是使用低频信号，因此会受到低频压缩机噪声的严重干扰（Francis et al.，2009）。有趣的是，与安静地点筑巢的鸟类相比，在嘈杂地点筑巢的鸟类的捕食率更低，繁殖成功率更高；这种关系既存在于物种内部，也存在筑巢群落的所有物种中。研究区的主要巢穴捕食者——西部灌丛鸦加利福尼亚黄斑鸦（*Aphelocoma Californica*）——不太可能出现在嘈杂的地点。这项研究表明，人为噪声不仅会影响某些物种的局部多度，而且还会影响物种之间的相互作用，从而对嘈杂地区的某些种群产生间接的积极影响。

城市的人工夜间照明以多种方式改变动物的行为（Longcore，Rich，2004；Eisenbeis，Hänel，2009；Kempenaers et al.，2010）。然而，对人工夜间照明的生物效应研究大多聚焦于个体，而其对种群影响的研究比较有限（Gaston et al.，2015）。人工照明增加可以将动物白天的活动延长至夜间，如觅食和鸣叫。觅食时间增加可以提升动物（如某些爬行动物）的成活率，在灯光下觅食可能反过来会使它们面临更高的被捕食率（Powell，Henderson，2008）。相反，在光线充足的城市环境中，喜爱在黑暗条件下觅食或猎捕的动物处于劣势（Longcore，Rich，2004）。一些快飞的蝙蝠被灯光和聚集在那里的大量昆虫猎物所吸引，而慢飞的物种则可能呈现相反的趋势（Rydell，Racey，1995）。欧洲的小菊头蝠（*Rhinolophus hipposideros*）飞行缓慢，易被白天活动的鸟类捕食。最近的一项实验发现，高压钠灯的安装减少了这个物种沿飞行路线的活动，并且延迟了其夜间飞行活动（Stone et al.，2009）。

人工照明可使需要黑暗条件的动物迷失方向，例如：迁徙的鸟类和昆虫，或者正在筑巢和孵化的乌龟（Witherington，Bjornal，1991；Longcore，Rich，2004；Gauthreaux，Belser，2006；New，2007；Bourgeois et al.，2009）。在阴暗的夜晚，当无法获得诸如月亮或星星之类的视觉线索时，人工照明会干扰鸟类的注意力，而白光和红光会干扰它们的磁感系统（Poot et al.，2008）。据估计，每年有数百计的迁徙鸟类受到人工夜间照明的影响，其中许多鸟类无法存活——它们可能会与发光物体发生碰撞，或被困在有灯光的区域继续绕圈飞行直到筋疲力尽（Poot et al.，2008；Eisenbeis，Hänel，2009）。

一项对加蓬的蓬加拉国家公园的棱皮龟的研究发现，刚从海滩上的巢穴孵化出来的幼龟偏向人造光爬行，而不是直奔大海（Bourgeois et al.，2009）。在加蓬首都利伯维尔灯光照射最多的海滩区域，这种影响最为明显，且几乎没有植被和原木等向陆地的形状线索。偏离直线路径会耗费大量的精力，花费额外的时间才能到达海洋会增加幼龟被捕食的风险（Bourgeois et al.，2009）。被人工灯光吸引的昆虫可能大量灭绝，这对种群甚至整个物种的持续性都有影响（Eisenbeis，Hänel，2009；van Langevelde et al.，2011）。偏振光污染是一种特殊的光污染，它是指从黑暗、有光泽的人造表面（如道路、建筑材料和塑料布等）反射的光，以与从深色水体反射的光类似的方式发生偏振（Horváth et al.，2009）。许多动物，包括至少 300 种水生昆虫，都能感知偏振光，并利用偏振光定向到合适的水生栖息地；它们可能被错误地吸引到人造偏光表面上，而人造偏光表面就像一个强大的生态陷阱（Horváth et al.，2009；见专栏 3.2）。

专栏 3.2

种群汇和生态陷阱

在生态学中，源-汇模型描述了一个多斑块栖息地系统中的种群动态（Pulliam，1988），其中至少一个高质量斑块栖息地的出生数大于死亡数（种群增长为正），还有一块低质量斑块栖息地的死亡数高于出生数（种群增长为负数）。从理论上说，前一个（源）迁入的多余个体可以维持后一个（汇）的种群。因此，源种群是个体净迁出者，而一个汇种群是净迁入者。源-汇模型在变化的城市环境中是有趣的应用。例如，某一特定物种在城市地区存在一个明显稳定的种群，可能表明该物种已成功适应城市中的栖息地条件（例如，它是一个城市适应者或开拓者；详见 3.5 节）。然而，仔细检查发现该种群的死亡率高或繁殖率低，并且实际上是从城市边缘的源种群迁入而维持种群稳定的。在新西兰丹尼丁的最近一项猫对鸟类进行捕食的研究就发现这样一种情况：在所研究的全部 6 个物种中，被猫杀死的鸟的数量都是非常大的，并且超过了整个城市范围内灰水鸭（*Rhipidura fuliginosa*）和引入的欧歌鸫(*Turdus Philomelos*)的数量(van Heezik et al.，2010)。

生态陷阱(ecological traps)发生在环境快速变化的地方，导致生物选择劣质栖息地或者汇栖息地，而不是高质量栖息地(Dwernychuk，Boag，1972)。考虑到个体根据环境线索选择合适的栖息地，如果栖息地质量或线索的分布发生变化，从而无法指引其找到合适的栖息地，则个体可能会选择较差的栖息地，并降低存活率或繁殖成功率(Schlaepfer et al.，2002)。例如，水生昆虫被人工偏振光源吸引，可能导致它们进入生态陷阱，它们将卵产在光亮的水泥地板或黑色塑料板上，而不是产在水中，或者在废油池中死亡(Horváth et al.，2009)。生态陷阱影响到栖息地的选择，这只是一系列进化陷阱中的一种(Schlaepfer et al.，2002)。生物体依靠环境线索作出其他各种决定，例如：何时繁殖、与谁交配、何时迁徙，或者从冬眠中苏醒。与城镇化相关的城市环境变化可能导致各种各样的进化陷阱，这些进化陷阱可以与其他更直接的城镇化的影响协同作用，从而危害种群和物种。

3.3 生物引进和入侵

3.3.1 植物和真菌

城市环境的特点是存在非本土物种。尽管这些物种大多数是有意或无意地被人引进的，但也有一些物种是在没有得到帮助的情况下进入城镇的，以开发它们特定的生态位和资源。世界各地的城市都有引进的植物，包括玫瑰、山茶花、水仙花、蓝花楹、叶子花、茉莉和鸡蛋花等园林植物，食用植物如西红柿、南瓜和食用草药，以及许多源自花园和农业的杂草。在城市中，植物多样性高的格局中，引进的植物占大部分(Porter et al.，2001；Hope et al.，2003；Thompson et al.，2003；Zipperer，Guntenspergen，2009)。而原先在该地区生存的本土植物，可能会在城镇化之后局部或在全球范围内灭绝(Thompson，Jones，1999；Williams et al.，2005；Kuhn，Klotz，2006；Hahs et al.，2009)。一些外来园林植物必须精心照料，才能在它们原产地的自然地理和气候范围之外生存，而另外一些外来园林植物则很少受到关注。后一类植物非常适应当地条件，在干扰之后能够拓殖新区域，对干扰具有很高的耐受性，会产生大量可以被风或动物传播的种子，或者可以无性繁殖，它们最有可能从花园新宠演变为归化种群(Baker，1974；Buist et al.，2000；Lake，Leishman，2004)。然而，一些因素会影响某个物种的归化种群是否具有入侵性，包括停留时间、气候、传播特征和繁殖压力(Richardson，Pyšek，2012)。

本土植物残留区域、公园和花园、工业区(包括在利用的和废弃的)和闲置土地，这些区域都容易受到外来植物和本地野生物种的入侵。最近一项对欧洲外来植物区系的研究发现，在所调查的10种栖息地中，工业栖息地以及管理的公园、花园支持了最大数量的引进植物物种(Lambdon et al.，2008)。土壤受到机械干扰，人为添加氮磷元素以及人为主导的景观破碎化，促进了外来物种入侵城市区域(和其他地方一样)(Riley，Banks，1996；Cilliers et al.，2008)。在城市栖息地中外来真菌也很常见，随地膜或作为本地植物的共生体被引入(Newbound，2008)。

城市溪流、池塘和湿地经常滋生非本土物种，这些非本土水生物种是人有意从花园池塘或鱼缸中引入的，或作为种子或植物物质从上游城市集水区被冲入水体中。Hussner(2009)发现，欧洲的4种水生杂草具有相对较高的生长率，而较高的营养水平又进一步提高了其生长速率，如城市排水沟中常见的那些杂草。这4种水生杂草可以无性繁殖，如水樱草(*Ludwigia grandiflora*)、狐尾藻(*Myriophyllum aquaticum*)，则可以从单叶中再生出新植物(Hussner，2009)。根据生长形式的不同，水草可以在城市溪流、池塘和湖泊表面形成致密的漂浮垫，从而取代原生水生植物，并降低水中氧气含量(如漂浮的毛毛蕨)。或者水草可在水体基质中茂密生长，从而排除了本土物种(如 *Typha* spp.；Ruiz-Avila，Klemm，1996；Zedler，Kercher，2004；Hussner，2009)。因此，水草在城市水道高养分条件下生长更快，在空间上排挤或遮蔽其他植物并大量繁殖，从而胜过本土物种。

互花米草(*Spartina alterniflora*)是一种来自北美洲的入侵物种，被有意引进到中国上海附近长江三角洲的岛屿上，以增加潮汐滩涂和底泥上的沉积物的堆积，并加速岛屿的扩张(Chen et al.，2008)。上海地区的人口和经济增长增加了对用于农业和住房的土地需求,而长江上游大型水坝项目的建设减少了长江口的泥沙量。但是,引入的米草竞争能力强,很快超过了本土植物,如海芦苇(*Scirpus mariqueter*)和常见的芦苇(*Phragmites australis*)。而有大量海芦苇的湿地,为水禽和迁徙候鸟提供了有效的栖息地,并且被拉姆塞尔湿地公约(RAMSAR)确认为具有国际重要性的湿地(Chen et al.,2008)。在长江九段沙岛的一项研究发现,所有鸟类(栖息鸟、浅水觅食鸟、水面鸭、海鸥)都避开了米草生长的区域(Ma et al.,2007)。尽管被中国国家环境保护总局列为有害入侵物种,但是仍在长江河口种植了米草(Ma et al.,2007)。对新土地的需求以及由米草主导的群落固碳能力优势(与本土植物群落相比),具有强有力的经济诱因,使人们继续采用这种做法(Chen et al.,2008)。

3.3.2 动物

除了引入各类植物和真菌外,城市中还经常有大量引进的动物。在城市区域,哺乳动物捕食者如猫、狗和狐狸等,几乎无处不在;许多引进的壁虎紧贴房屋墙壁;城市池塘、溪流中许多异国的鱼和海龟也随处可见;逃跑的家养鹦鹉的后代,在附近公园的树林中鸣叫;外来的蚯蚓在城市森林中的土壤中挖洞;而非本土海洋动物,如多毛虫、苔藓虫、海星、贻贝也常栖息在城市的港口(Steinberg et al.,1997;Hewitt et al.,2004;Beckerman et al.,2007)。例如,在澳大利亚南部的菲利普湾港,发现了72种被引进的动物,以及29种来源不明的物种。其中许多种动物是附着在从遥远的国际港口驶来的船只的船体或压载水中而引进的(Hewitt et al.,2004)。在城市环境中,一些引进动物以低密度存在,对本地物种或生态系统没有明显的影响。相反,其他一些物种则通过争夺资源、直接捕食或更复杂的过程(如捕食者-介导竞争)而与本土物种产生强烈的相互作用。

世界上许多城市都大量引进鸟类,包括画眉、椋鸟、麻雀和岩鸽。原产于亚洲的家八哥(*Acridotheres tristis*),被引入澳大利亚、新西兰、南非、中东、欧洲和北美,以及包括夏威夷和马达加斯加在内的许多岛屿(Peacock et al.,2007;BirdLife International,2012)。在郊区和城市自然保护区,它的种群密度常常很高,澳大利亚堪培拉记录的种群密度大于100只/km^2(Pell,Tidemann,1997a)。

它具有强烈的领地性，被自然保护联盟(IUCN)列为世界上最差的100种入侵物种之一(Lowe et al.，2000)。家八哥在树洞(包括人工巢箱)中筑巢共同栖息，并与本土鸟类争夺栖息和筑巢地点(Pell，Tidemann，1997b)。它们取代了其他成对繁殖的鸟类，会从树洞中抛出这些繁殖鸟类的巢和幼鸟(Pell，Tidemann，1997b；Blanvillain et al.，2003)。因此，在城市地区，它们很可能会减少天然或人工本地鸟类的繁殖成功率和种群数量(Pell，Tidemann，1997b；Blanvillain et al.，2003)。

引进的捕食者可以在城市陆地和水生生境中达到很高的种群密度，并且可能会对它们所捕食的本地物种的种群数量产生重大影响。例如，最近的一项调查估计，在5个月内英国有900万只家猫捕获并带回了9200万只猎物(95%置信区间：85%～100%)，其中包括5700万只哺乳动物、2700万只鸟类和500万只爬行动物(Woods et al.，2003)。虽然这些数量很大，但只相当于每只猫每两周捕获一个猎物。记录的猎物包括20种哺乳动物和44种鸟类，其中一些是值得保护的，如水鼩鼱和黄颈鼠。项圈上系着铃铛的猫和夜间待在室内的猫往往捕获较少的哺乳动物，但后者白天在户外捕获了更多的爬行动物和两栖动物(Woods et al.，2003；图3.6)。最近在英国布里斯托的一项研究试图量化猫捕食对城市鸟类种群的影响(Baker et al.，2008)。尽管在准确地测量鸟类种群密度和繁殖率以及猫捕获的猎物个体数量方面存在困难，但家猫的捕食可能会对常见城市鸟类的种群产生重大影响，如林岩鹨(*Prunella modularz*)、欧洲知更鸟(*Erithacus rubecula*)和鹪鹩(*Troglodutes troglodytes*)。在4个研究区中，这些物种被杀死的个体数量估计超过了幼雏的数量(Baker et al.，2008)。在这种情况下，很可能通过引进来维持城市鸟类的数量，而城市地区成了种群的汇入地(见专栏3.2)。

引进的捕食者还可能导致城市环境中呈现更复杂的种间相互作用。Beckerman等(2007)的模型指出，英国城市地区，猫的高种群密度足以通过低水平的捕食和对繁殖力的压制作用来减少鸟类数量。由于猫的存在，而引起鸟类行为改变甚至导致繁殖力小幅度下降，也可以解释过去30年来观察到的城市常见鸟类物种急剧减少的现象(Beckerman et al.，2007)。赤狐是另一种捕食者，已被广泛引入其本土范围之外的地区，在城市地区普遍存在。该物种于1885年被首次引入美国加利福尼亚州，但其分布一直受到限制，直到1970年开始迅速扩张

3.3 生物引进和入侵

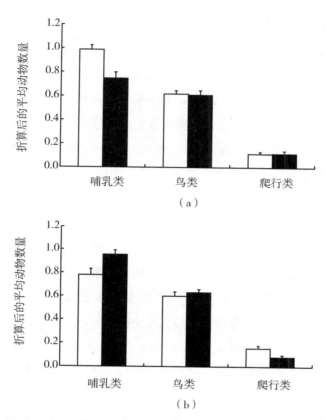

图 3.6 英国 282 只家猫捕获的猎物：哺乳类、鸟类和爬行类动物。(a) 带铃铛的猫 (黑柱) 和不带铃铛的猫 (白柱)；(b) 晚上允许出入的猫 (黑柱) 和不允许出入的猫 (白柱) (Woods et al., 2003。图 5 和图 6) (注：数据为取对数后的平均值)

(Lewis et al., 1999)。20 世纪 90 年代，在奥勒冈县，赤狐 (*Vulpes vulpes*) 非常普遍，这里的城市发展减少了适合本土捕食者的栖息地，同时又建造了公园、高尔夫球场和墓地等为狐狸提供庇护的开敞空间。Lewis 等 (1999) 将此作为中型捕食者释放到城市的实例 (Soulé et al., 1988)，城镇化进程减少了该食物链系统中顶级捕食者 [郊狼 (*Canis latrans*) 和美洲狮 (*Felis concolor*)] 的可利用栖息地，从而允许较低级别的中型捕食者如引入的赤狐变得异常多 (有关城市环境中营养关系的进一步讨论，请参见专栏 4.2)。赤狐是杂食性捕食者，被认为是对加利福尼亚州一些较小本土动物的威胁，其中包括濒临灭绝的长嘴秧鸡 (*Rallus longirostrus levipes*)和圣华金沙狐(*Vulpes macrotis mutica*)(Lewis et al., 1999)。

引入的淡水鱼,如鲤鱼、金鱼和食蚊鱼在城市池塘、湖泊和溪流中很常见,它们以本地无脊椎动物、鱼类和两栖动物幼虫为食(Lowe et al., 2000; Hamer, McDonnell, 2008)。为了减少蚊幼虫和蚊媒疾病的发病率,全世界多地从北美东部引进了两种食蚊鱼(*Gambusia affinis*, *G. holbrooki*)(Pyke, 2008)。食蚊鱼具有很广的环境耐受性,它可以生活在各种有静止或缓慢流动水的水生环境中且水温在 0~45℃ 之间都能适应(Cherry et al., 1976; Pyke, 2008)。它们寿命短,可以高密度出现并快速繁殖,但是其在控制蚊子方面的效果尚不确定[见文献 Pyke (2008)中的讨论]。由于它们在某些城市水生生境中的营养关系位置居高不下,且种群多度高,食蚊鱼(*Gambusia* spp.)可以在若干营养水平上影响多种本土类群。通过直接捕食,食蚊鱼可以减少或消灭其他小鱼、两栖动物幼虫和无脊椎动物的种群;结果,这些被减少或消灭的物种的竞争者或猎物可能间接受益(Pyke, 2008)。例如:Jassby 等(1977a, 1977b)在实验室中观察到食蚊鱼对大型浮游动物的选择性捕食,导致对浮游植物的捕食压力降低,增加了细菌、浮游植物和轮虫的多度。

城市区域高密度引进物种除了破坏食物网外,还可能导致栖息地结构发生变化,这会对其他物种造成不利影响。北太平洋海星(*Asterias amurensis*)是一种软沉积物栖息地中的泛化捕食者,于 1986 年在澳大利亚司马尼亚州的德文特河口被首次发现(Buttermore et al., 1994)。此后,它已成为该系统中主要的无脊椎动物捕食者,密度≤46 个/m^2,被认为是对海洋底栖群落和商业贝类渔业的主要威胁(Ross et al., 2002)。一项控制性实验表明,它对天然双壳类动物施加了巨大的捕食压力。10 周后,在北太平洋海星自然密度的情况下,薄皮蛤(*Fulvia tenuicostata*)的幼贝从 300 只/m^2 减少到 35 只/m^2;在网箱内的北太平洋海星密度为 1/m^2 时,薄皮蛤的幼贝的平均密度只有 17 只/m^2(Ross et al., 2002)。相反,在排除北太平洋海星的情况下,薄皮蛤的平均尺寸显著增加。塔斯马尼亚的斑点手鱼(*Brachionichthys hirsutus*)濒临灭绝也暗示了此海星的存在,它要么直接捕食斑点手鱼的卵块,要么捕食斑点手鱼的产卵基质海鞘(Bruce, Green, 1998)。斑点手鱼是一条细小的、缓慢移动的底栖鱼类,用类似于人手的胸鳍和腹鳍行走(图 3.7)。它的分布非常有限,20 世纪 80 年代,数量下降与德温特河口北太平洋海星种群的建立与扩大相吻合(Bruce，Green，1998)。

图3.7 濒临灭绝的斑点手鱼是澳大利亚塔斯马尼亚德温特河口特有物种。该物种受到北太平洋海星的威胁(https://commons.wikimedia.org/wiki/)

3.4 人为干扰

城市中的居民以多种方式干扰定居在城市中的其他生物：他们收集真菌、动植物作为食物或在市场上出售；将它们从一个地方移到另一个地方；挖出他们不喜欢的植物；修剪或践踏植物；收集岩石和伐倒树木，破坏了无脊椎动物和脊椎动物的栖息地；改变多种动物的行为。一般而言，城市区域人为干扰强度随着人口密度的增加而增加，但也随着文化习俗、社会态度和社区的社会经济地位的变化而变化。食用真菌、动植物以及可以用作燃料的耸立或倒下的木材，很可能在贫困地区比在富裕地区承受更大的来自人类的压力(参见Stoian，2005)。

城市中残存植被经常受到人类践踏，造成的后果是土壤压实、根茎生长减少，植物覆盖度降低以及地表层或林下层物种丰富度的下降（Florgård，2009）。踩踏对底层植被的影响随土壤类型、土壤深度和湿度水平以及植被群落的组成和结构而变化（Florgård，2009；Hill，Pickering，2009）。在澳大利亚黄金海岸的一个亚热带城市保护区进行的踩踏实验表明，具有蕨类林下植被的地方对踩踏的抵抗力很弱，经过10次通行后，蕨类的相对高度和覆盖度降低了（Hill，Pickering，2009）。以丛生草和引入的草坪草为主导的林下植被具有中、高的抗践踏性。人为践踏对植物和土壤造成的物理损害会阻碍植物的生长，杀死单株植物，并降低成功补充植物幼株的可能性。因此，践踏可能会影响存活率，以致某些植物的种群被排除在践踏场所之外（图3.1）。相反，耐受这种人为干扰的植物物种，可能会在沿着踩踏的区域旺盛生长（Hamberg et al.，2008）。

潮间带的动植物被踩踏在经常有人走动的沿海地区很普遍，对不同物种的覆盖率和密度有不同的影响（Keough，Quinn，1998）。随着时间的流逝，这可能导致易于被反复践踏物种的局部灭绝。另一种人为干扰，即收集岩石用于城市景观花园建设，与澳大利亚东南部阔头蛇（*Hoploaphalus bungaroides*）的衰退有关（Webb，Shine，2000）。移除这些岩石，会减少蛇及其重要猎物——绒壁虎（*Oedura lesueurii*）的日间活动点。在新威尔士州南部的莫顿国家公园进行的一项实验表明，混凝土摊铺机可以为这些爬行动物在退化的地区提供人工休憩场所（Webb，Shine，2000）。现在从新南威尔士州的公共土地上收集灌木石是非法的，根据《1995年新南威尔士州受威胁物种保护法》，清除灌木石被列为关键的威胁过程。

在城市地区，无论有没有家犬，一种更广泛、危害程度较小的人类活动是散步。关于人类遛狗对本地野生动植物，特别是鸟类的潜在影响已有相当多的争论。人们发现，马德里的城市公园中行人（没有狗）的干扰会降低鸟类的多样性和数量，以及单个物种出现在特定公园中的可能性（Fernández-Juricic，2000）。在澳大利亚悉尼市区边缘的90个林地上进行了一项实验性，与没有人走的对照组相比，研究人行走对鸟类的影响（Banks，Bryant，2007）。在人和狗走过去后的10分钟内，人遛狗地点的鸟类种类比对照点少35%，鸟类数量比对照点减少41%（Banks，Bryant，2007）。仅有人行走的场所的鸟类多样性和多度，也比对照点低，但影响却比遛狗的场地小。在地面活动的鸟类似乎最容易受遛狗行为的影响，比对照点的鸟类数量减少一半。有趣的是，即使在经常遛狗的地方，鸟类也会受到狗的影响，这表明当地鸟类没有习惯于持续的干扰（Banks，Bryant，2007）。在穿过城市森林的小径上骑自行车也可能打扰鸟类和其他野生动植物，导致种群密度下降（George，Crooks，2006；Thompson，2015）。

英国最近的一项研究发现，人类散步及其狗对林莺（*Sylvia undata*）有重要影响（Murison et al.，2007）。这种鸟在英国的分布有限，仅在英格兰南部的沿海荒地上繁殖。英格兰的许多沿海灌木丛都在城市中心附近，经常有人想带着狗看鸟和欣赏自然风景（Murison et al.，2007；Underhill-Day，Liley，2007）。该研究表明，人类访客及其狗的干扰将林莺繁殖期延迟 6 周，导致每年育雏数量减少（Murison et al.，2007）。有人提出了两种机制来解释这种现象：干扰延迟了孵化日期，导致雏鸟的生长时期与最佳无脊椎动物猎物密度的时期不一致，从而降低了雏鸟的存活率；干扰事件直接打断了成年鸟的觅食和对雏鸟的饲喂（Murison et al.，2007）。

人为干扰海洋栖息地的一个例子：在加拿大不列颠哥伦比亚省的约翰斯顿海峡，人们发现船只（商业捕鱼船和观鲸船）的出现，改变了虎鲸（*Orlinus orca*）的行为（Williams et al.，2006）。鲸鱼在光滑卵石滩上觅食和摩擦身体的时间减少了，据估计失去觅食机会使它们的能量摄入减少了 18%。在靠近城市的海洋栖息地中，大量船只通行可能会影响一系列鲸类动物（参见 Dans et al.，2008；Stockin et al.，2008），从而导致种群数量下降，因为干扰频繁，严重影响它们的进食活动。

3.5 对城市环境中种群的随机效应

如第 2 章所述，已知三种随机性（环境的随机性、种群随机性和灾难性）会影响种群中个体的出生率和死亡率（Lande，1993）。在城市内或城市郊区的栖息地碎片中，孤立的本土物种种群易受环境和种群随机性的影响（Bolger et al.，1997；Kéry et al.，2003；Sattler et al.，2010）。环境条件的随机波动会降低出生率和（或）增加死亡率，这可能足以使这些种群灭绝。同样，一个或多个繁殖季节中繁殖失败和（或）低于平均成年个体存活率，可能会导致少数种群的局部灭绝，而这种隔离也降低了将来成功重新定居的可能性。无论种群大小，都可能受到环境灾难的影响，例如：野火、洪水、疾病、强风暴和长期干旱。尽管可以通过管理燃料负荷、主动灭火和对水路进行结构改造，来降低城市中心区发生特定灾难的风险，例如野火和洪水，但是不能将其完全排除（参见 Keeley et al.，1999；Nyambod，2010；Buxton et al.，2011）。城市地区生境破碎化的好处之一是减少了诸如火灾、疾病等在孤立种群中蔓延的可能性，因此在发生灾难时，这些种群同时灭绝的可能性较小（Fahrig，2003）。一般而言，确定性和随机性过程会协同影响城市环境中种群的增长率，以及随时间推移而持续或灭绝的可能性。

3.6　总结

种群（及其所属的物种）以多种方式应对城镇化。城镇化过程可以分为三大类：生物物理过程、生物引进和人为干扰。这些过程改变了生物赖以生存的资源数量和质量，以及它们在时间和空间上的分布。它们还可以通过微妙而戏剧性的方式改变物种内部和物种之间的相互作用。这些变化影响着四个重要因素：出生率、死亡率、迁出和迁入，进而可能改变系统内、种内和种间的相互作用。因此，城镇化进程最终可能导致城市建设之前存在的种群和（或）物种局部灭绝，其他物种的多度降低，可以在城市环境中蓬勃发展的某些本土和外来物种的多度增加，并且可以改变物种的全球分布。在下一章中，我们将讨论生态群落对城镇化的不同响应。

思考题

1. 描述构成种群动态的四个过程或生命率。
2. 城镇化的生物物理过程是通过什么机制和途径影响这些种群水平的？
3. 什么是集合种群？城市中集合种群是如何运作的？
4. 城市噪声和光照环境对本土物种有什么重要影响？
5. 什么是生态陷阱？你将如何识别生态陷阱？
6. 人为干扰和生物引进如何影响城市环境中的本土物种和种群？

本章参考文献

Agneta M, Burton S. (1990) Terrestrial and aquatic bryophytes as monitors of environmental contaminants in urban and industrial habitats. *Botanical Journal of the Linnean Society*, 104, 267-280.

Amorim M C P, Vasconcelos R O. (2008) Variability in the mating calls of the Lusitanian toadfish *Halobatrachus didactylus*: cues for potential individual recognition. *Journal of Fish Biology*, 73, 1267-1283.

ANZECC. (2000) *Australian and New Zealand Guidelines for Fresh and Marine Water Quality Volume 1, The Guidelines*. Australian and New Zealand Environment and Conservation Council and theAgriculture and Resource Management Council of Australia and New Zealand, Australian Government, Canberra.

Asfaw Z, Tadesse M. (2001) Prospects for sustainable use and development of wild food plants in Ethiopia. *Economic Botany*, 55, 47-62.

Baker H G. (1974) The evolution of weeds. *Annual Review of Ecology and Systematics*, 5, 1-24.

Baker L A, Brazel A J, Selover N, et al. (2002) Urbanization and warming of Phoenix (Ari-zona, USA): Impacts, feedbacks and mitigation. *Urban Ecosystems*, 6, 183-203.

Baker P J, Molony S E, Stone E, et al. (2008) Cats about town: is predation by free-ranging pet cats *Felis catus* likely to affect urban bird populations? *Ibis*, 150, 86-99.

Banks P B, Bryant J V. (2007) Four-legged friend or foe? Dog walking displaces native birds from natural areas. *Biology Letters*, 3, 611-613.

Banks W A, Altmann J, Sapolsky R M, et al. (2003) Serum leptin as a marker for a syndrome X-like condition in wild baboons. *The Journal of Clinical Endocrinology and Metabolism*, 88, 1234-1240.

Barber J R, Crooks K R, Fristrup K M. (2010) The costs of chronic noise exposure for terrestrial organisms. *Trends in Ecology and Evolution*, 25, 180-189.

Batary P, Baldi A. (2004) Evidence of an edge effect on avian nest success. *Conservation Biology*, 18, 389-400.

Bates A J, Sadler J P, Fairbrass A J, et al. (2011) Changing bee and hoverfly pollinator assemblages along as urban-rural gradient. *PLoS ONE* 6, e23459. doi: 10.1371/journal.pone.0023459.

Baur B, Baur A. (1993) Climatic warming due to thermal-radiation from an urban area as possible cause for the local extinction of a land snail. *Journal of Applied Ecology*, 30, 333-340.

Beale C M, Monaghan P. (2004) Human disturbance: people as predation-free predators? *Journal of Applied Ecology*, 41, 335-343.

Beckerman A P, Boots M, Gaston K J. (2007) Urban bird declines and the fear of cats. *Animal Conservation*, 10, 320-325.

Beckmann J P, Berger J. (2003) Rapid ecological and behavioural changes in carnivores: the response of black bears (*Ursus americanus*) to altered food. *Journal of Zoology*, 261, 207-212.

Bee M A, Swanson E M. (2007) Auditory masking of anuran advertisement calls by road traffic noise. *Animal Behaviour*, 74, 1765-1776.

Beebee T J. (1979) Habitats of the British amphibians (2): suburban parks and gardens. *Biological Conservation*, 15, 241-257.

Bermúdez-Cuamatzin E, Rios-Chelen A A, Gil D, et al. (2009) Strategies of song adaptation to urban noise in the house finch: syllable pitch plasticity of differential syllable use? *Behaviour*, 146, 1269-1286.

Beyene A, Legesse W, Triest L, et al. (2009) Urban impact on ecological integrity of nearby rivers in developing countries: the Borkena River in highland Ethiopia. *Environmental Monitoring and Assessment*, 153, 461-476.

Bhattacharya M, Primack R B, Gerwein J. (2003) Are roads and railroads barriers to bumblebee movement in a temperate suburban conservation area? *Biological Conservation*, 109, 37-45.

BirdLife International. (2012) *Acridotheres tristis*. The IUCN Red List of Threatened Species. Version 2015.2. www.iucnredlist.org (accessed on 20/07/2015).

Blanvillain C, Salducci J M, Tuturural G, et al. (2003) Impact of introduced birds on the recovery of the Tahiti Flycatcher (*Pomarea nigra*), a critically endangered forest bird of Tahiti. *Biological Conservation*, 109, 197-205.

Blickley J L, Blackwood D L, Patricelli G L. (2012a) Experimental evidence for the effects of chronic anthropogenic noise on abundance of greater sage-grouse at leks. *Conservation Biology*, 26, 461-471.

Blickley J L, Word K, Krakauer A H, et al. (2012b) The effect of experimental exposure to chronic noise on fecal corticosteroid metabolites in lekking male greater sage-grouse (*Centrocercus urophasianus*). *PLoS ONE* 7 (11): e50462. doi: 10.1371/journal.pone.0050462.

Bolger D T, Alberts A C, Sauvajot R M, et al. (1997) Response of rodents to habitat fragmentation in coastal southern California. *Ecological Applications*, 7, 552-563.

Booker D J. (2003) Hydraulic modelling of fish habitat in urban rivers during high flows. *Hydrological Processes*, 17, 577-599.

Both C, van Asch M, Bijlsma R G, et al. (2008) Climate change and unequal phenological changes across four trophic levels: constraints or adaptations? *Journal of Animal Ecology*, 78, 73-83.

Bourgeois S, Gilot-Fromont E, Viallefont A, et al. (2009) Influence of artificial lights, logs, and erosion on leatherback sea turtle hatchling orientation at Pongara National Park, Gabon. *Biological Conversation*, 142, 85-93.

Brattstrom B H, Bondello M C. (1983) Effects of off-road vehicle noise on desert vertebrates. In *Environmental Effects of Off-road Vehicles*. Webb R H, Wilshire H H eds. Springer-Verlag, New York, 167-206.

Bruce B A, Green M A. (1998) *The Spotted Handfish 1999-2001 Recovery Plan*. Australian Government, Canberra. http://www.environment.gov.au/resource/spotted-handfish-1999-2001-recovery-plan.

Brumm H. (2004) The impact of environmental noise on song amplitude in a territorial bird. *Journal of Animal Ecology*, 73, 434-440.

Buist M, Yates C J, Ladd P G. (2000) Ecological characteristics of *Brachychiton populneus* (Sterculiaceae) (kurrajong) in relation to the invasion of urban bushland in south-western Australia. *Austral Ecology*, 25, 487-496.

Bunnell F L. (1999) What habitat is an island? In *Forest Fragmentation: Wildlife and Management Implications*. Rochelle J A, Lehmann L A, Wisniewski J eds. Brill, Leiden, 1-31.

Burgess J, Harrison C M, Limb M. (1988) People, parks, and the urban green: a study of popular meanings and values for open spaces in the city. *Urban Studies*, 25, 455-473.

Buttermore R E, Turner E, Morrice M G. (1994) The introduced northern Pacific seastar *Asterias amurensis* in Tasmania. *Memoirs of the Queensland Museum*, 36, 21-25.

Buxton M, Haynes R, Mercer D, et al. (2011) Vulnerability to bushfire risk at Melbourne's urban fringe: the failure of regulatory land use planning. *Geographical Research*, 49, 1-12.

Chamberlain D E, Cannon A R, Toms M P, et al. (2009) Avian productivity in urban landscapes: a review and meta-analysis. *Ibis*, 151, 1-18.

Chen J, Zhao B, Ren W, et al. (2008) Invasive *Spartina* and reduced sediments: Shanghai's dangerous silver bullet. *Journal of Plant Ecology*, 1, 79-84.

Cherry D S, Rodgers J H Jr, Cairnes J Jr, et al. (1976) Responses of mosquitofish (*Gambusia affinis*) to ash effluent and thermal stress. *Transactions of the American Fisheries Society*, 105, 686-694.

Cilliers S S, Williams N S G, Barnard F J. (2008) Patterns of exotic plant invasions in fragmented urban and rural grasslands across continents. *Landscape Ecology*, 23, 1243-1256.

Clark T N, Lloyd R, Wong K K, et al. (2002) Amenities drive urban growth. *Journal of Urban Affairs*, 24, 493-515.

Clarke G P, White P C L, Harris S. (1998) Effects of roads on badger *Meles meles* populations in southwest England. *Biological Conservation*, 86, 117-124.

Close D C, Messina G, Krauss S L, et al. (2006) Conservation biology of the rare species *Conospermum undulatum* and *Macarthuria keigheryi* in an urban bushland remnant. *Australian Journal of Botany*, 54, 583-593.

Contesse P, Hegglin D, Gloor S, et al. (2004) The diet of urban foxes (*Vulpes vulpes*) and the availability of anthropogenic food in the city of Zurich, Switzerland. *Mammalian Biology*, 69, 81-95.

Czech B, Krausman P R. (1997) Distribution and causation of species endangerment in the United States. *Science*, 277, 1116.

Czech B, Krausman P R, Devers P K. (2000) Economic associations among causes of species endangerment in the United States. *BioScience*, 50, 593-601.

Dans S L, Crespo E A, Pedraza S N, et al. (2008) Dusky dolphin and tourist interaction: effect on diurnal feeding behaviour. *Marine Ecology Progress Series*, 369, 287-296.

De la Puente D, Ochoa C, Viejo J L. (2008) Butterflies killed on roads (Lepidoptera, Papilionoidea) in "El Regajal-Mar de Ontigola" Nature Reserve (Aranjuez, Spain). *XVII Bienal de la Real Sociedad Española de Historia Natural*, 17, 137-152.

Dos Santos M E, Modesto T, Matos R J, et al. (2000) Sound production by the Lusitanian toad fish, Halobatrachus didactylus. *The International Journal of Animal Sound and its Recording*, 10, 309-321.

Drinnan I N. (2005) The search for fragmentation thresholds in a southern Sydney suburb. *Biological Conversation*, 124, 339-349.

Durant J M, Hjermann D Ø, Ottersen G, et al. (2007) Climate and the match or mismatch between predator requirements and resource availability. *Climate Research*, 33, 271-283.

Dwernychuk L W, Boag D A. (1972) Ducks nesting in association with gulls-an ecological trap? *Canadian Journal of Zoology*, 50, 559-563.

Eisenbeis G, Hänel A. (2009) Light pollution and the impact of artificial night lighting on insects. In *Ecology of Cities and Towns*. McDonnell M J, Hahs A K, Breuste J H eds. Cambridge University Press, Cambridge, 243-263.

Erbe C. (2002) Underwater noise of whale-watching boats and potential effects on killer whales (*Orcinus orca*), based on an acoustic impact model. *Marine Mammal Science*,

18, 394-418.

Fahrig L. (2003) Effects of habitat fragmentation on biodiversity. *Annual Review of Ecology, Evolution and Systematics*, 34, 487-515.

Faulkner S. (2004) Urbanization impacts on the structure and function of forest wetlands. *Urban Ecosystems*, 7, 89-106.

Fernández-Juricic E. (2000) Avifaunal use of wooded streets in an urban landscape. *Conservation Biology*, 14, 513-521.

Fischer J, Lindenmayer D B. (2006) Beyond fragmentation: the continuum model for fauna research and conservation in human-modified landscapes. *Oikos*, 112, 473-480.

Florgård C. (2009) Preservation of original natural vegetation in urban areas: an overview. In *Ecology of Cities and Towns: A Comparative Approach*. McDonnell M J, Hahs A K, Breuste J H eds. Cambridge University Press, Cambridge, 380-398.

Foster E, Curtis L R, Gundersen D. (2014) Toxic Contaminants in the urban aquatic environment. In *Wild Salmonids in the Urbanizing Pacific Northwest*. Yeakley J A, Maas-Hebner K G, Hughes R M eds. Springer, New York, 123-144.

Fox C H. (2006) Coyotes and humans: can we coexist? *Proceedings of the Vertebrate Pest Conference*, 22, 287-293.

Francis C D, Ortega C P, Cruz A. (2009) Noise pollution changes avian communities and species interactions. *Current Biology*, 19, 1-5.

Frankham R, Ballou J D, Briscoe D A. (2002) *Introduction to Conservation Genetics*. Cambridge University Press, Cambridge.

Frankham R, Ballou J D, Briscoe D A. (2004) *A Primer of Conservation Genetics*. Cambridge University Press, Cambridge.

Fuller R A, Irvine K N, Devine-Wright P, et al. (2007) Psychological benefits of greenspace increase with biodiversity. *Biology Letters*, 3, 390-394.

Gadsdon S R, Dagley J R, Wolseley P A, et al. (2010) Relationships between lichen community composition and concentrations of NO_2 and NH_3. *Environmental Pollution*, 158, 2553-2560.

Garland T. (1984) Physiological correlates of locomotory performance in a lizard: an allometric approach. *American Journal of Physiology*, 247, 806-815.

Gaston K J, Visser M E, Hölker F. (2015) The biological impacts of artificial light at night: the research challenge. *Philosophical Transactions of the Royal Society B* 370, 20140133.

Gates E, Gysel L W. (1978) Avian nest dispersion and fledging success in field-forest ecotones. *Ecology*, 59, 871-883.

Gauthreaux S A, Belser C G. (2006) Effects of artificial night lighting on migrating birds. In *Ecological Consequences of Artificial Night Lighting*. Rich C, Longcore T eds. Island Press, Washington DC, 67-93.

Gehrt S D. (2004) Ecology and management of striped skunks, raccoons, and coyotes in urban landscapes. In *Predators and People: From Conflict to Conservation*. Fascione N, Delach A, Smith M eds. Island Press, Washington DC, 81-104.

George S L, Crooks K R. (2006) Recreations and large mammal activity in an urban nature reserve. *Biological Conservation*, 133, 107-117.

Gibbs J P, Shriver W G. (2002) Estimating the effects of road mortality on turtle populations. *Conservation Biology*, 16, 1647-1652.

Gilbert O L. (1969) The effect of SO_2 on lichens and bryophytes around Newcastle upon Tyne. In *Air Pollution: Proceedings of the First European Congress on the Influence of Air Pollution on Plants and Animals*. Center for Agricultural Publishing and Documentation, Wageningen: 223-235.

Gilbert O L. (1971) Some indirect effects of air pollution on bark-living invertebrates. *Journal of Applied Ecology*, 8, 77-84.

Giordani P. (2007) Is the diversity of epiphytic lichens a reliable indicator of air pollution? A case study from Italy. *Environmental Pollution*, 146, 317-323.

Grimm N B, Foster D, Groffman P, et al. (2008) The changing landscape: ecosystem responses to urbanization and pollution across climate and societal gradients. *Frontiers in Ecology and the Environment*, 6, 264-272.

Hahs A K, McDonnell M J, McMarthy M A, et al. (2009) A global synthesis of plant extinction rates in urban areas. *Ecology Letters*, 12, 1165-1173.

Hale J M, Heard G W, Smith K L, et al. (2013) Structure and fragmentation of growling grass frog metapopulations. *Conservation Genetics*, 14, 313-322.

Halfwerk W, Slabbekoorn H. (2009) A behavioural mechanism explaining noise-dependent frequency use in urban birdsong. *Animal Behaviour*, 78, 1301-1307.

Halfwerk W, Holleman L J M, Lessells C M, et al. (2011) Negative impact of traffic noise on avian reproductive success. *Journal of Applied Ecology*, 48, 210-219.

Hamberg L, Lehvävirta S, Malmivaara-Lämsä M, et al. (2008) The effects of habitat edges and trampling on understorey vegetation in urban forests in Helsinki, Finland. *Applied Vegetation Science*, 11, 83-86.

Hamer A J, McDonnell M J. (2008) Amphibian ecology and conservation in the urbanising world: a review. *Biological Conservation*, 141, 2432-2449.

Hanski I A. (1994) A practical model of metapopulation dynamics. *Journal of Animal Ecology*, 63, 151-162.

Hanski I A. (1998) Metapopulation dynamics. *Nature*, 396, 41-49.

Hanski I A, Ovaskainen O. (2002) Extinction debt at extinction threshold. *Conservation Biology*, 16, 666-673.

Harper M J. (2005) Home range and den use of common brushtail possums (*Trichosurus vulpec-ula*) in urban forest remnants. *Wildlife Research*, 32, 681-687.

Harper M J, McCarthy M A, van der Ree R. (2008) Resources at the landscape scale influence possum abundance. *Austral Ecology*, 33, 243-252.

Hawksworth D L. (1970) Lichens as litmus for air pollution: a historical review. *International Journal of Environmental Studies*, 1, 281-296.

Heard G W, McCarthy M A, Scroggie M P, et al. (2013) A Bayesian model of metapopulation viability, with application to an endangered amphibian. *Diversity and Distributions*, 19, 555-566.

Hedrick P W, Kalinowski S T. (2000) Inbreeding depression in conservation biology. *Annual Review of Ecology and Systematics*, 31, 139.

Hels T, Buchwald E. (2001) The effect of road kills on amphibian populations.

Biological Conservation, 99, 331-340.

Hewitt C L, Campbell M L, Thresher R E, et al. (2004) Introduced and cryptogenic species in Port Phillip Bay, Victoria, Australia. *Marine Biology*, 144, 183-202.

Hill R, Pickering C. (2009) Difference in resistance of three subtropical vegetation types to experimental trampling. *Journal of Environmental Management*, 90, 1305-1312.

Hitchings S P, Beebee T J C. (1998) Loss of genetic diversity and fitness in Common Toad (*Bufo bufo*) populations isolated by inimical habitat. *Journal of Evolutionary Biology*, 11, 269-283.

Hope D Gries C, Zhu W, et al. (2003) Socioeconomics drive urban plant diversity. *Proceedings of the National Academy of Sciences of the United States of America*, 100, 8788-8792.

Horváth G, Kriska G, Malik P, et al. (2009) Polarized light pollution: A new kind of ecological photopollution. *Frontiers in Ecology and the Environment*, 7, 317-325.

How R A, Dell J. (2000) Ground vertebrate fauna of Perth's vegetation remnants: impacts of 170 years of urbanization. *Pacific Conservation Biology*, 6, 198-217.

Hussner A. (2009) Growth and photosynthesis of four invasive aquatic plant species in Europe. *Weed Research*, 49, 506-515.

Jassby A, Dudzik M, Rees J, et al. (1977a) *Production cycles in aquatic microcosms*. U. S. Environmental Protection Agency Report EPA-600/7-77-077. Environmental Protection Agency, Washington DC.

Jassby A, Rees J, Dudzik M, et al. (1977b) *Trophic structure modifications by planktivorous fish in aquatic microcosms*. U. S. Environmental Protection Agency Report EPA-600/7-77-096. Envi-ronmental Protection Agency, Washington DC.

Johnson P T J, Hoverman J T, McKenzie V J, et al. (2013) Urbanization and wetland communities: applying metacommunity theory to understanding the local and landscape effects. *Journal of Applied Ecology*, 50, 34-42.

Keeley J E, Fotheringham C J, Morais M. (1999) Re-examining fire suppression impacts on brushland fire regimes. *Science*, 284, 1829-1832.

Keller L F, Waller D M. (2002) Inbreeding effects in wild populations. *Trends in Ecology and Evolution*, 17, 230-241.

Kempenaers B, Borgstrom P, Loes P, et al. (2010) Artificial night lighting affects dawn song, extrapair siring success, and lay date in songbirds. *Current Biology*, 20, 1735-1739.

Kentula M E, Gwin S E, Pierson S M. (2004) Tracking changes in wetlands with urbanization: sixteen years of experience in Portland, Oregon, USA. *Wetlands*, 24, 734-743.

Keough M J, Quinn G P. (1998) Effects of periodic disturbances from trampling on rocky intertidal algal beds. *Ecological Applications*, 8, 141-161.

Kéry M, Matthies D, Schmid B. (2003) Demographic stochasticity in population fragments of the declining distylous perennial. *Primula veris* (Primulaceae). *Basic and Applied Ecology*, 4, 197-206.

Kight C R, Swaddle J P. (2011) How and why environmental noise impacts animals: an integrative, mechanistic review. *Ecology Letters*, 14, 1052-1061.

King S A, Buckney R T. (2000) Urbanization and exotic plants in northern Sydney streams.

Konrad C P, Booth D B. (2005) Hydrologic changes in urban streams and their ecological significance. *American Fisheries Society Symposium*, 47, 157-177.

Kuhn I, Klotz S. (2006) Urbanization and homogenization-comparing the floras of urban and rural areas in Germany. *Biological Conservation*, 127, 292-300.

Kuussaari M, Bommarco R, Heikkinen R K, et al. (2009) Extinction debt: a challenge for biodiversity conservation. *Trends in Ecology and Evolution*, 24, 564-571.

Lake J C, Leishman M R. (2004) Invasion success of exotic plants in natural ecosystems: the role of disturbance, plant attributes, and freedom from herbivores. *Biological Conservation*, 117, 215-226.

Lambdon P W, Pysek P, Basnou C, et al. (2008) Alien flora of Europe: species diversity, temporal trends, geographical patterns, and research needs. *Preslia*, 80,

101-149.

Lampe U, Reinhold K, Schmoll T. (2014) How grasshoppers respond to road noise: developmental plasticity and population differentiation in acoustic signalling. *Functional Ecology*, 28, 660-668.

Lampe U, Schmoll T, Franzke A, et al. (2012) Staying tuned: grasshoppers from noisy roadside habitats produce courtship signals with elevated frequency components. *Functional Ecology*, 26, 1348-1354.

Lande R. (1993) Risks of population extinction from demographic and environmental stochasticity and random catastrophes. *The American Naturalist*, 142, 911-927.

Lättman H, Bergman K, Rapp M, et al. (2014) Decline in lichen biodiversity on oak trunks due to urbanization. *Nordic Journal of Botany*, 32, 518-528.

Leblanc S C F, DeSloover J. (1970) Relation between industrialization and the distribution and growth of epiphytic lichens and mosses in Montreal. *Canadian Journal of Botany*, 48, 1485-1496.

Lenten L J A, Moosa I A. (2003) An empirical investigation into long-term climate change in Australia. *Environmental Modelling and Software*, 18, 59-70.

Lesage V, Barrette C, Kingsley M C S, et al. (1999) The effect of vessel noise on the vocal behaviour of belugas in the St. Lawrence River estuary, Canada. *Marine Mammal Science*, 15, 65-84.

Levins R. (1969) Some demographic and genetic consequences of environmental heterogeneity for biological control. *Bulletin of the Ecological Society of America*, 15, 237-240.

Lewis J C, Sallee K L, Golightly R T Jr. (1999) Introduction and range expansion of nonnative red foxes (*Vulpes vulpes*) in California. *American Midland Naturalist*, 142, 372-381.

Lindenmayer D B, Franklin J F. (2002) *Conserving Forest Biodiversity: A Comprehensive Multi-scaled Approach*. Island Press, Washington DC.

Longcore T, Rich C. (2004) Ecological light pollution. *Frontiers in Ecology and the Environment*, 2, 191-198.

Lowe S, Browne M, Boudjelas S, et al. (2000) 100 *of the World's Worst Invasive Alien Species—A selection from the Global Invasive Species Database.* Invasive Species Specialist Group, SpeciesSurvival Commission, International Union for the Conservation of Nature (IUCN): Gland, Switzerland. http://www.issg.org/database/species/reference_files/100English.pdf(accessed 10/10/2013).

Luniak M. (2004) Synurbization-adaptation of animal wildlife to urban development. In *Proceedings 4th International Symposium on Urban Wildlife Conservation.* Shaw W W, Harris L K, Vandruff L eds. School of Natural Resources, University of Arizona, Tucson, 50-55.

Ma Z, Gan X, Choi C, et al. (2007) Wintering bird communities in newly-formed wetland in the Yangtze River estuary. *Ecological Research*, 22, 115-124.

Maharjan M R. (1998) The flow and distribution of costs and benefits in the Chuliban community forest, Dhankuta district, Nepal. *Rural Development Forestry Network*, *Network Paper* 23e, Summer 1998. http://www.odi.org/sites/odi.org.uk/files/odi-assets/publications-opinion-files/1186.pdf(accessed 18/07/2015)

Marchand M N, Litvaitis J A. (2004) Effects of habitat features and landscape composition on the population structure of a common aquatic turtle in a region undergoing rapid development. *Conservation Biology*, 18, 758-767.

McCarthy M A, Lindenmayer D B, Drechsler M. (1997) Extinction debts and risks faced by abundant species. *Conservation Biology*, 11, 221-226.

McCauley R D, Fewtrell J, Popper A N. (2003) High intensity anthropogenic sound damages fish ears. *Journal of the Acoustical Society of America*, 113, 638-642.

McClenaghan L R, Truesdale H D. (2002) Genetic structure of endangered Stephens' kangaroo rat populations in southern California. *Southwestern Naturalist*, 47, 539-549.

McClure C J W, Ware H E, Carlisle J, et al. (2013) An experimental investigation into the effects of traffic noise on distributions of birds: avoiding the phantom road. *Proceedings of the Royal Society B* 280, 20132290. 10.1098/rspb.2013.2290

McDonald R I, Kareiva P, Forman R T T. (2008) The implications of current and future urbanization for global protected areas and biodiversity conservation. *Conservation Biology*, 141, 1695-1703.

McKenna D, McKenna K, Malcom S B, et al. (2001) Mortality of lepidoptera along roadways in Central Illinois. *Journal of the Lepidopterists' Society*, 55, 63-68.

McKinney M L. (2002) Urbanization, biodiversity, and conservation. *BioScience*, 52, 883-890.

Menzel A. (2003) Plant phenological anomalies in Germany and their relation to air temperature and NAO. *Climate Change*, 57, 243-263.

Milner-Gulland E J, Bennett E L, the SCB 2002 Annual Meeting Wild Meat Group (2003) Wild meat: The bigger picture. *Trends in Ecology and Evolution* 18, 351-357.

Mitchell J C, Brown R E J, Bartholomew B. (2008) Urban Herpetology. In *Herpetological Conservation*, *Number* 3. Society for the Study of Amphibians and Reptiles, Salt Lake City.

Montgomery J. (1998) Making a city: Urbanity, vitality, and urban design. *Journal of Urban Design*, 3, 93-116.

Morgan J W. (1998) Patterns of invasion of an urban remnant of a species-rich grassland in southeastern Australia by non-native plant species. *Journal of Vegetation Science*, 9, 181-190.

Morrissey C A, Stanton D W G, Tyler C R, et al. (2014) Developmental impairment in Eurasian dipper nestlings exposed to urban stream pollutants. *Environmental Toxicology and Chemistry*, 33, 1315-1323.

Muerk C D, Zvyagna N, Gardner R O, et al. (2009) Environmental, social and spatial determinants of urban arboreal character in Auckland, New Zealand. In *Ecology of Cities and Towns: a Comparative Approach*. McDonnell M J, Hahs A K, Breuste J H eds. Cambridge University Press, Cambridge, 287-307.

Muñoz P T, Torres F P, Megías A G. (2015) Effects of roads on insects: a review. *Biodiversity and Conservation*, 24, 659-682.

Murison G, Bullock J M, Underhill-Day J, et al. (2007) Habitat type determines the effects of disturbance on the breeding productivity of the Dartford Warbler *Sylvia undata*. *Ibis*, 149, 16-26.

New T R. (2007) Politicians, poison, and moths: ambiguity over the icon status of the Bogong moth(*Agrotis infusa*) (Noctuidae) in Australia. *Journal of Insect Conservation*, 11, 219-220.

Newbound M, McCarthy M A, Lebel T. (2010) Fungi and the urban environment: A review. *Landscape and Urban Planning*, 96, 138-145.

Newbound M G. 2008. Fungal diversity in remnant vegetation patches along an urban to rural gradient. PhD Thesis, University of Melbourne, Melbourne.

Noël S, Ouellet M, Galois P, et al. (2007) Impact of urban fragmentation on the genetic structure of the eastern red-backed salamander. *Conservation Genetics*, 8, 599-606.

Nyambod E M. (2010) Environmental consequences of rapid urbanisation: Bamenda City, Cameroon. *Journal of Environmental Protection*, 1, 15-23.

Parks S E, Clark C W, Tyack P L. (2007) Short and long-term changes in right whale calling behavior: The potential effects of noise on acoustic communication. *Journal of the Acoustical Society of America*, 122, 3725-3731.

Parris K M. (2015) Ecological impacts of road noise and options for mitigation. In *Ecology of Roads: A Practitioners' Guide to Impacts and Mitigation*. van der Ree R, Grilo C, Smith D eds. Wiley-Blackwell, New York, 151-158.

Parris K M, Hazell D L. (2005) Biotic effects of climate change in urban environments: the case of the grey-headed flying-fox (*Pteropus poliocephalus*) in Melbourne, Australia. *Biological Conservation*, 124, 267-276.

Parris K M, Schneider A. (2009) Impacts of traffic noise and traffic volume on birds of roadside habitats. *Ecology and Society* 14, 29. http://www.ecologyandsociety.org/vol14/iss1/art29/.

Parris K M, Velik-Lord M, North J M A. 2009. Frogs call at a higher pitch in traffic noise. *Ecology and Society* 14, 25. http://www.ecologyandsociety.org/vol14/iss1/art25/.

Partecke J, Van't Hof T J, Gwinner E. (2005) Underlying physiological control of reproduction in urban and forest-dwelling European blackbirds *Turdus merula*. *Journal of Avian Biology*, 36, 295-305.

Peacock D S, Van Rensburg B J, Robertson M P. (2007) The distribution and spread of the invasive alien common myna, *Acridotheres tristis* L. (Aves: Sturnidae), in southern Africa. *South African Journal of Science*, 103, 465-473.

Pell A S, Tidemann C R. (1997a) Theecology of the common Myna in urban nature reserves in the Australian Capital Territory. *EMU*, 97, 141-149.

Pell A S, Tidemann C R. (1997b) The impact of two exotic hollow-nesting birds on two native parrots in savannah and woodland in eastern Australia. *Biological Conservation*, 79, 145-153.

Peres C A, Palacios E. (2007) Basin-wide effects of game harvest on vertebrate population densities in Amazonian forests: implications for animal-mediated seed dispersal. *Biotropica*, 39, 304-315.

Poot H, Ens B J, de Vries H, et al. (2008) Green light for nocturnally migrating birds. *Ecology and Society* 13, 47. http://www.ecologyandsociety.org/vol13/iss2/art47/.

Popper A N, Hastings M C. (2009) The effects of anthropogenic sources of sound on fishes. *Journal of Fish Biology*, 75, 455-489.

Porter E E, Forschner B R, Blair R B. (2001) Woody vegetation and canopy fragmentation along a forest-to-urban gradient. *Urban Ecosystems*, 5, 131-151.

Potvin D A, Parris K M, Mulder R A. (2011) Geographically pervasive effects of urban noise on frequency and syllable rate of songs and calls in silvereyes(*Zosterops lateralis*). *Proceedings of the Royal Society B*, 278, 2464-2469.

Potvin D A, Mulder R A. (2013) Immediate, independent adjustment of call pitch and amplitude in response to varying background noise by silvereyes(*Zosterops lateralis*). *Behavioral Ecology*, 24, 1363-1368.

Powell R, Henderson R W. 2008. Urban herpetology in the West Indies. In Mitchell J C, Jung Brown R E, Bartholomew B eds. *Urban Herpetology. Herpetological*

Conservation vol. 3, Society for the Study of Amphibians and Reptiles, Salt Lake City, 389-404.

Puckett K J, Nieboer E, Flora W P, et al. (1973) Sulphur dioxide: its effect on photo-synthetic ^{14}C fixation in lichens and suggested mechanisms of phytotoxicity. *New Phytologist*, 72, 141-154.

Pulliam H R. (1988) Sources, sinks, and population regulation. *American Naturalist*, 132, 652-661.

Pyke G H. (2008) Plague minnow or mosquito fish? A review of the biology and impacts of introduced *Gambusia* species. *Annual Review of Ecology, Evolution, and Systematics*, 39, 171-191.

Pyle R M. (1995) A history of Lepidoptera conservation, with special reference to its Remingtonian debt. *Journal of the Lepidopterists' Society*, 49, 397-411.

Ranta P. (2001) Changes in urban lichen diversity after a fall in sulphur dioxide levels in the city of Tampere, SW Finland. *Annales Botanici Fennici*, 38, 295-304.

Rao R S P, Girish M K S. (2007) Road kills: assessing insect casualties using flagship taxon. *Current Science*, 92, 830-837.

Rebele F. (1994) Ecology and special features of urban ecosystems. *Global Ecology and Biogeography Letters*, 4, 173-187.

Reed D H, Frankham R. (2003) Correlation between fitness and genetic diversity. *Conservation Biology*, 17, 230-237.

Reijnen R, Foppen R. (1994) The effects of car traffic on breeding bird populations in woodland. I. Evidence of reduced habitat quality for Willow Warblers (*Phylloscopus trochilus*) breeding close to a highway. *Journal of Applied Ecology*, 31, 85-94.

Reijnen R, Foppen R, Meeuwsen H. (1996) The effects of traffic on the density of breeding birds in Dutch agricultural grasslands. *Biological Conservation*, 75, 255-260.

Reijnen R, Foppen R, ter Braak C, et al. (1995) The effects of car traffic on breeding bird populations in woodland. III. *Reduction of density in relation to the proximity of main roads. Journal of Applied Ecology*, 32, 187-202.

Resh V H, Jackson J K. (1993) Rapid assessment approaches to biomonitoring using benthic macroinvertebrates. In *Freshwater Biomonitoring and Benthic Macroinvertebrates*. Rosenberg D M, Resh V H eds. Chapman and Hall, New York, 195-233.

Richardson D M, Pyšek P. (2012) Naturalization of introduced plants: ecological drivers of biogeographical patterns. *New Phytologist*, 196, 383-396.

Riley S J, Banks R G. (1996) The role of phosphorus and heavy metals in the spread of weeds in urban bushland: An example from the Lane Cove Valley, NSW, Australia. *The Science of the Total Environment*, 182, 39-52.

Riley S P D, Pollinger J P, Sauvajot R M, et al. (2006) A southern California freeway is a physical and social barrier to gene flow in carnivores. *Molecular Ecology*, 15, 1733-1742.

Riley S P D, Sauvajot R M, Fuller T K, et al. (2003) Effects of urbanization and habitat fragmentation on bobcats and coyotes in southern California. *Conservation Biology*, 17, 566-576.

Robbins C T, Schwartz C C, Felicetti L A. (2004) Nutritional ecology of ursids: a review of newer methods and management implications. *Ursus*, 15, 161-171.

Robinson N A, Marks C A. (2001) Genetic structure and dispersal of red foxes (*Vulpes vulpes*) in urban Melbourne. *Australian Journal of Zoology*, 49, 589-601.

Roetzer T, Wittenzeller M, Haeckel H, et al. (2000) Phenology in central Europe-differences and trends of spring phenophases in urban and rural areas. *International Journal of Biometeorology*, 44, 60-66.

Ross D J, Johnson C R, Hewitt C L. (2002) Impact of introducing seastars *Asterias amurensis* on survivorship of juvenile commercial bivalves *Fulvia tenuicostata*. *Marine Ecology Progress Series*, 241, 99-112.

Rubbo M J, Kiesecker J M. (2005) Amphibian breeding distribution in an urbanized landscape. *Conservation Biology*, 19, 504-511.

Ruiz-Avila R J, Klemm V V. (1996) Management of *Hydrocotyle ranunculoides* L. f, an aquatic invasive weed of urban waterways in Western Australia. *Hydrobiologia*, 340, 187-190.

Rydell J, Racey P A. (1995) Street lamps and the feeding ecology of insectivorous bats. *Symposium of the Zoological Society of London*, 67, 291-307.

Samuel Y, Morreale S J, Clark C W, et al. (2005) Underwater, low-frequency noise in a coastal sea turtle habitat. *The Journal of the Acoustical Society of America*, 117, 1465-1472.

Sattler T, Borcard D, Alettaz R, et al. (2010) Spider, bee, and bird communities in cities are shaped by environmental control and high stochasticity. *Ecology*, 91, 3343-3353.

Sauvajot R M, Buechner M, Kamradt D A, et al. (1998) Patterns of human disturbance and response by small mammals and birds in chaparral near urban development. *Urban Ecosystems*, 2, 279-297.

Scheifele P M, Andrew S, Cooper R A, et al. (2005) Indication of a Lombard vocal response in the St. Lawrence River beluga. *The Journal of Acoustic Society of America*, 117, 1486-1492.

Schlaepfer M A, Runge M C, Sherman P W. (2002) Ecological and evolutionary traps. *Trends in Ecology and Evolution*, 17, 474-480.

Seiler A, Helldin J, Seiler C. (2004) Road mortality in Swedish mammals: Results of a drivers' questionnaire. *Wildlife Biology*, 10, 225-233.

Sharpe D M, Stearns F, Leitner L A, et al. (1986) Fate of natural vegetation during urban development of rural landscapes in Southeastern Wisconsin. *Urban Ecology*, 9, 267-287.

Shannon G, Angeloni L M, Wittemyer G, et al. (2014) Road traffic noise modifies behaviour of a keystone species. *Animal Behaviour*, 94, 135-141.

Shochat E, Warren P S, Faeth S H. (2006) Future directions in urban ecology. *Trends in Ecology and Evolution*, 21, 661-662.

Singer F J. (1978) Behavior of mountain goats in relation to U. S. Highway 2, Glacier National Park, Montana. *Journal of Wildlife Management*, 42, 591-597.

Skórka P, Lenda M, Moron D, et al. (2013) Factors affecting road mortality and the suitability of road verges for butterflies. *Biological Conservation*, 159, 148-157.

Slabbekoorn H, Peet M. (2003) Birds sing at a higher pitch in urban noise. *Nature*, 424, 267.

Slabbekoorn H, Bouton N, van Opzeeland I, et al. (2010) A noisy spring: The impact of globally rising underwater sound levels on fish. *Trends in Ecology and Evolution*, 25, 419-427.

Snodgrass J W, Casey R E, Joseph D, et al. (2008) Microcosm investigations of stormwater pond sediment toxicity to embryonic and larval amphibians: Variation in sensitivity among species. *Environmental Pollution*, 154, 291-297.

Soulé M E, Alberts A C, Bolger D T. (1992) The effects of habitat fragmentation on chaparral plants and vertebrates. *Oikos*, 63, 39-47.

Soulé M E, Bolger D T, Alberts A C, et al. (1988) Reconstructing dynamics of rapid extinction of chaparral-requiring birds in urban habitat island. *Conservation Biology*, 2, 75-92.

Spinks P Q, Pauly G B, Crayon J J, et al. (2003) Survival of the western pond turtle (*Emys marmorata*) in an urban California environment. *Biological Conservation*, 113, 257-267.

Steinberg D A, Pouyat R V, Parmelee R W, et al. (1997) Earthworm abundance and nitrogen mineralization rates along an urban-rural land use gradient. *Soil Biology and Biochemistry*, 29, 427-430.

Stockin K A, Lusseau D, Binedell V, et al. (2008) Tourism affects the behavioural budget of the common dolphin *Delphinus* sp. in the Hauraki Gulf, New Zealand. *Marine Ecology Progress Series*, 355, 287-295.

Stoian D. (2005) Making the best of two worlds: rural and peri-urban livelihood options sustained by non-timber forest products from the Bolivian Amazon. *World Development*, 33, 1473-1490.

Stone E L, Jones G, Harris S. (2009) Street lighting disturbs commuting bats. *Current Biology*, 19, 1123-1127.

Storper M, Manville M. (2006) Behaviour, preferences, and cities: Urban theory and urban resurgence. *Urban Studies*, 43, 1247-1274.

Sun J W C, Narins P M. (2005) Anthropogenic sounds differentially affect amphibian call rate. *Biological Conservation*, 121, 419-427.

Temple S A. (1987) Predation on turtle nests increases near ecological edges. *Copeia*, 1987, 250-252.

Thompson B. (2015) Recreational trails reduce the density of ground-dwelling birds in protected areas. *Environmental Management*, 55, 1181-1190.

Thompson K, Austin K C, Smith R M, et al. (2003) Urban domestic gardens(I): putting small-scale plant diversity in context. *Journal of Vegetation Science*, 14, 71-78.

Thompson K, Jones A. (1999) Human population density and prediction on local plant extinction in Britain. *Conservation Biology*, 13, 185-189.

Tilman D, May R M, Lehman C L, et al. (1994) Habitat destruction and the extinction debt. *Nature*, 371, 65-66.

Torok S J, Morris C J G, Skinner C, et al. (2001) Urban heat island features of southeast Australian towns. *Australian Meteorological Magazine*, 50, 1-13.

Torok S J, Nicholls N. (1996) A historical annual temperature dataset for Australia. *Australian Meteorological Magazine*, 45, 251-260.

Trombulak S C, Frissell C A. (2000) Review of ecological effects of roads on terrestrial and aquatic communities. *Conservation Biology*, 14, 18-30.

Underhill-Day J C, Liley D. (2007) Visitor patterns on southern heaths: A review of visitor access patterns to heathlands in the UK and the relevance to Annex I bird species. *Ibis*, 149, 112-119.

van der Ree R, McDonnell M J, Temby I D, et al. (2006) The establishment and dynamics of a recently established urban camp of Pteropus poliocephalus outside their geographic range. *Journal of Zoology*, 268, 177-185.

van Heezik Y, Smyth A, Adams A, et al. (2010) Do domestic cats impose an unsustainable harvest on urban bird populations? *Biological Conservation*, 143, 121-130.

van Langevelde F, Ettema J A, Donners M, et al. (2011) Effects of spectral composition of artificial light on the attraction of moths. *Biological Conservation*, 144,

2274-2281.

Vandergast A G, Bohonak A J, Weissman D B, et al. (2007) Understanding the genetic effects of recent habitat fragmentation in the context of evolutionary history: phylogeography and landscape genetics of a southern California endemic Jerusalem cricket (Orthoptera: Stenopel-matidae: *Stenopelmatus*). *Molecular Ecology*, 16, 977-992.

Vasconcelos R O, Amorim M C P, Ladich F. (2007) Effects of ship noise on the detectability of communication signals in the Lusitanian toadfish. *Journal of Experimental Biology*, 210, 2104-2112.

Vasconcelos R O, Carriço R, Ramoa A, et al. (2012) Vocal behavior predicts reproductive success in a teleost fish. *Behavioral Ecology*, 23, 375-383.

Vasconcelos R O, Simões J M, Almada V C, et al. (2010) Vocal behaviour during territorial intrusions in the Lusitanian toadfish: boatwhistles also function as territorial "keep-out" signals. *Ethology*, 116, 155-165.

Walsh C J, Roy A H, Feminella J W, et al. (2005) The urban stream syndrome: current knowledge and the search for a cure. *Journal of North American Benthological Society*, 24, 706-723.

Walsh C J, Waller K A, Gehling J, et al. (2007) Riverine invertebrate assemblages are degraded more by catchment urbanisation than by riparian deforestation. *Freshwater Biology*, 52, 574-587.

Warren P S, Katti M, Ermann M, et al. (2006) Urban bioacoustics: it's not just noise. *Animal Behaviour*, 71, 491-502.

Webb J K, Shine R. (2000) Paving the way for habitat restoration: can artificial rocks restore degraded habitats of endangered reptiles? *Biological Conservation*, 92, 93-99.

Wellborn G A, Skelly D K, Werner E E. (1996) Mechanisms creating community structure across a freshwater habitat gradient. *Annual Review of Ecology and Systematics*, 27, 337-363.

White M A, Nemani R R, Thornton P E, et al. (2002) Satellite evidence of phenological differences between urbanized and rural areas of the Eastern United

States broadleaf forest. *Ecosystems*, 5, 260-277.

Williams N S G, Schwartz M W, Vesk P A, et al. (2009) A conceptual framework for predicting the effects of urban environments on floras. *Journal of Ecology*, 97, 4-9.

Williams N S G, Morgan J W, McDonnell M J, et al. (2005) Plant traits and local extinctions in natural grasslands along an urban-rural gradient. *Journal of Ecology*, 93, 1203-1213.

Williams R, Lusseau D, Hammond P S. (2006) Estimating relative energetic costs of human disturbance to killer whales (*Orcinus orca*). *Biological Conservation*, 133, 301-311.

Witherington B E, Bjorndal K A. (1991) Influences of artificial lighting on the seaward orientation of hatchling Loggerhead Turtles *Caretta caretta*. *Biological Conservation*, 55, 139-149.

Wood W E, Yezerinac S M. (2006) Song Sparrow (*Melospiza melodia*) song varies with urban noise. *The Auk*, 123, 650-659.

Woods M, McDonald R A, Harris S. (2003) Predation of wildlife by domestic cats *Felis catus* in Great Britain. *Mammal Review*, 33, 174-188.

Yule C M, Gan J Y, Jinggut T, et al. (2015) Urbanization affects food webs and leaf-litter decomposition in a tropical stream in Malaysia. *Freshwater Science*, 34, 702-715.

Zedler J B, Kercher S. (2004) Causes and consequences of invasive plants in wetlands: Opportunities, opportunists, and outcomes. *Critical Reviews in Plant Sciences*, 23, 431-452.

Zhou L, Dickinson R E, Tian Y, et al. (2004) Evidence for a significant urbanization effect on climate in China. *Proceedings of the National Academy of Sciences of the United States of America*, 101, 9640-9544.

Zipperer W C, Guntenspergen G R. (2009) Vegetation composition and structure of forest patches along urban-rural gradients. In *Ecology of Cities and Towns: A Comparative Approach*. McDonnell M J, Hahs A K, Breuste J H eds. Cambridge University Press, Cambridge, 274-286.

第4章 群落响应城镇化

4.1 概述

现在，我们将注意力转移到生态群落如何响应城镇化这一问题上。首次建设城镇时，有时会全部破坏本地生态群落。城市发展速度、强度和（或）规模越大，这种破坏越彻底。然而，某些物种的种群将在城镇化的初始阶段幸存下来，随着时间的流逝，与其他物种（包括本土物种和引进物种）的种群将通过扩散和物种形成结合在一起。因此，在生境丧失对群落造成直接影响之后，通常可以在种群或物种层面观察到城镇化的影响。然而，种群和物种对城镇化进程的不同反应，可能导致生态群落发生重要改变和出现新群落模式。城市栖息地的特殊性，可以导致新型生态群落的形成，其中一些在自然环境中没有明显的类似情况。城镇化对城市以外的陆地、淡水和海洋生态群落也有重大影响。

在这里，我们将生态群落定义为一类在时间和空间上同时发生潜在相互作用的物种的集合。这就是所谓的利己主义的群落概念（Gleason，1917，1926）。可以将它与有机群落概念进行对比（Clements，1916，1936），有机群落概念强调了在一个较长的时期内，作为一个群落共存的物种之间的相互作用。因此，它们形成的群落被视为"超有机体"。群落的个体论概念，使我们可以考虑任何时空尺度上的生态群落（Vellend，2010），例如：从生活在城市公园的一棵树上的无脊椎动物群落，或生活在砖墙缝隙中的植物群落，到生活在整个城市或世界范围每个城市中的无脊椎动物群落或植物群落。每个群落的研究时间范围，可能从一天到数百年或数千年。

群落生态学家志在研究生态群落内物种的多样性、多度和组成，以及影响群落结构的潜在过程和相互作用（见专栏4.1）。在过去50年里，物种共存以及生态群落中观察到的物种丰富度和多度的模式的理论基础受到了广泛关注，导致出现各种各样的概念、假设和理论（Palmer，1994；Vellend，2010）。针对这种扩散，许多生态学家认为群落生态学缺乏一种适用于所有（或至少大多数）生态系统和环境的通用理论（参见 Palmer，1994；Lawton，1999；Simberloff，2004）。Vellend（2010）提出一个对现有群落理论进行分类的框架，着重于四个群落层次过程——选择、生态漂移、扩散和物种形成——类似于群落遗传学中的选择、遗传漂移、基因流和突变四个过程。

专栏4.1

监测生态群落

生态学家以多种方式描述群落，例如通过其所包含的物种数量（物种丰富度），这些物种的鉴别（组成）以及它们在时间和空间上的变化（如β多样性、系统发育多样性、稳定性、流通量、继承性和弹性）。群落中不同物种的相对多度，即是一个群落中的个体比例，这些信息可以使用多样性或均匀性指标进行汇总，例如 Shannon-Wiener 多样性指数（Begossi，1996；Spellerberg，Fedor，2003）。一系列的理论和模型被用来预测非城市环境中生态群落的物种相对多度，包括对数正态分布（Preston，1948）、折断线模型（MacArthur，1960）、零和多样式分布（Hubbell，2001）。有趣的是，考虑在城市地区观察到的群落模式类型，是否与在自然区域观察到的相似，生态群落中物种相对多度的各种模型是否在所有环境中都做出可靠的预测？

对城市环境中生态群落物种丰富度的关注，可能会排除那里特定物种的重要信息。例如，郊区花园可能比城市边缘的同类草原、林地或沙漠支持更多的植物物种（请参阅3.3.1节）。但是，如果花园中的物种包括许多本地常见的、外来的物种和园艺物种，而原生植被群落包括稀有的、特有的物种时，则前者的生态和保护价值被认为低于后者。在更大空间尺度上，德国整个城市区域的维管束植物群落具有比农业或半自然区域更高的物种丰富度，但这并没有更高的系统发育多样性（Knapp et al.，2008）。城市植物群落的物种丰富度在很大程度上是由于城市区域中有大量密切相关、功能相似的物种。

生态群落的组成通常受物种之间的相互作用影响，包括对资源的竞争，如空间、营养、水和光（对于真菌和植物），或食物、水、庇护所和繁殖场地（对于动物）。其他可以影响群落构成的种间相互作用，包括捕食、互惠、寄生和捕食者介导的竞争（Polis et al.，1989；Hatcher et al.，2006）。如第3章所示，城镇化可以极大地改变时间和空间上的资源可用性。这些变化进而影响物种之间的相互作用，导致生态群落的多样性和组成发生变化。在世界各地不同城市中观察到的群落级变化的独特性或普遍性是城市生态学中的一个关键问题。

在群落生态学中，选择产生于物种之间的内在差异（物种之间的特征）和同一物种个体之间的内在差异（种内变异），这些差异会影响它们在给定环境中的适应性（Vellend，2010；Bolnick et al.，2011；Violle et al.，2012）。生态漂移是由群落内的随机性或不可预测性引起的；扩散是个体在地理空间中的移动；物种形成是随着时间的推移，新物种的进化（Vellend，2010）。例如：生态位理论强调选择，简单的中性理论强调生态漂移，集群理论考虑了选择、漂移和扩散。Vellend（2010）关于生态群落的一般理论考虑了所有四个过程：物种通过扩散被添加到群落中，然后它们的相对多度通过生态漂移、选择和进一步扩散而形成。Nemergut等（2013）对Vellend的方案做了轻微修改，该方案使用进化多样化而非物种形成作为第四群落层次过程。这超出了进化变化及其对群落动态的潜在影响的考虑范围，而不仅仅是在新物种产生的特殊情况下。它还包括传统物种概念可能不适合的生物的重要进化变化，如许多微生物（Konstantinidis et al.，2006；Burke et al.，2011；Nemergut et al.，2013）。

无论是一般还是具体，群落生态学中很少有理论是考虑到城市环境而发展起来的。那么，这些理论在城市范围内的适用程度如何？是否可以使用它们来帮助了解城镇化对生态群落的影响？它们能否预测城市生态群落的特征，如物种丰富度、物种组成、相对多度或时空动态？还是我们需要针对城市群落发展一个新的生态理论？在本章，我们将在研究城镇化对群落的一些关键影响的同时来回答这些问题。研究城镇化对生态群落的影响的一种方法,是思考城镇化如何改变四个群落层次的过程:选择、生态漂移、扩散和进化多样化。

4.2 选择:城市生态学中的生态位理论

4.2.1 生态位和环境梯度

在生态群落这一级,选择的过程取决于物种之间以及同一物种的个体之间的功能差异。我们可以将"选择"的概念应用于一个群落中的物种,而不是种群中的等位基因,通过将"种群"视为一个群落中所有物种的所有个体,且(在简单的情况下)将个体所属的物种作为主要利益特征。正如在一个种群中选择可能有利于一个等位基因而不是另一个等位基因一样,在一个群落中选择可能偏向一个物种的个体而不是另一个物种的个体(Velland,2010)。这些差异会影响每个物种与其环境以及所遇到的其他物种相互作用的方式。选择可以是恒定的(当一个物种中的个体适应性不变,并且物种之间的适应性以一致的方式变化时),也可以是密度制约(当一个物种中的个体适应性取决于该物种中个体的密度或群落中的其他物种)。无论是恒定的还是密度制约,选择还可以随时间和空间而变化(Velland,2010)。

例如,在特定环境条件下,一个物种的个体可能比其他物种的个体具有适应性优势,也就是说,它们更有可能生存和繁殖。当环境条件因地理空间而异时,不同物种会青睐不同的位置。观察到自然界中这些生态模式,提出了生态位概念。"生态位"(niche)一词,最早由 Grinnell(1917)在生态学中使用,后来由 Hutchinson(1957)将其定义为一个物种可以存在的多维环境空间(或超体积)。该空间的每个维度代表着与所讨论物种相关的环境变量或资源,被视为连续体或梯度。例如,让我们考虑两个环境变量:年平均气温和年平均降雨量。一种树可能在凉爽、潮湿的环境下生长,而另一种树则在较热、干燥的环境下生长。这两种树多在由两个气候变量定义的环境空间中所占的比例不同,每种树的多度预计将随着气候梯度变化而变化(图4.1)。界定物种生态位的环境空间,因所考虑的每个其他环境变量(如土壤深度、朝向、太阳辐射)而增加一个维度。

4.2 选择：城市生态学中的生态位理论

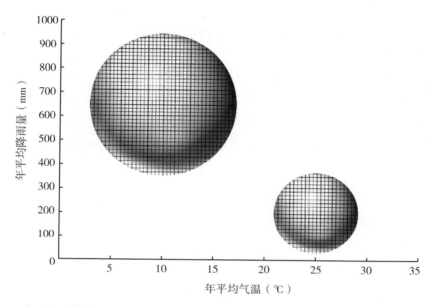

图 4.1 由两个环境变量(年平均气温和年平均降雨量)定义的两种树木的生态位示意图；每个物种的多度(用 3D 圆展示)从生态位边缘到中心逐渐增加。与右边的物种相比，左边的物种适应凉爽、潮湿的环境，并且具有更大的生态位，而右边的物种则适应更热、更干燥的环境

一个物种在没有其他物种的情况下，所能占据的整个环境空间被称为基础生态位。但是，一个物种在自然界中占据的实际环境空间被称为现实生态位，通常比基础生态位要小得多，因为多个物种的个体可能在特定的时间和地点寻求相同的资源。随着同城分布物种的基础生态位重叠度增加，我们预计它们之间的相互作用也会增加。竞争排斥原理，也称为高斯原理(Gause，1934；Hardin，1960)，是指具有相同基础生态位的两个物种不能无限期共存，因为一个物种最终必须通过迁移或引起另一物种局部灭绝的方式来排除另一个物种。尽管这一原理在自然、随机环境中很少被遵循，但它为描述性的和定量的群落生态学提供了重要的理论基础。生态位概念是群落生态学中一系列概念、理论和模型的核心，如梯度范式、所有类型的栖息地模型(包括那些包含功能特征的模型)、生态组团概念、资源竞争模型、捕食者-猎物模型、食物网和营养级联(专栏 4.2)。

103

> 专栏 4.2

城市栖息地中的食物网和营养相互作用

食物网描绘了生态群落中物种之间的营养相互作用。在群落中营养互动最常见的类型是谁吃了谁,但是营养互动还包括谁是寄生虫或寄主,谁与谁竞争,谁与谁是共生关系。具有相同捕食者和猎物的单个物种或物种组,或称为功能组,在食物网内形成节点;节点之间通过营养相联系,这些联系代表了它们之间的能量流(Thompson et al., 2012;图 4.2)。食物网的营养级别或位置很多,首先是腐生生物(分解有机物质的微生物)和自养生物(通过光合作用生产能量,如植物)位于底部,然后是食草动物(以植物为食的动物)、初级捕食者(吃食草动物的动物)、次级捕食者(吃初级捕食者的动物)和杂食动物(同时吃动物、植物的动物)。最后四类生物都是异养生物,因为它们是以其他生物作为食物。食物网的最高营养级别包含顶部或顶层捕食者,尽管它们经常具有各种寄生虫,但是它们本身没有捕食者(图 4.2)。虽然食物网比较复杂,但是它们往往具有某些特征,包括顶层捕食者、中间物种和基础物种的比例,以及网中的营养级数(通常 3~5 级;Pimm et al., 1991)。

顶层捕食者的多度和活动似乎强烈影响某些食物网的结构和功能(自上而下的控制)。相比之下,基础物种的多度或对系统的营养输入似乎可以调节其他食物网(自下而上的控制)。在实践中,大多数食物网可能包含自上而下和自下而上的控制元素(Pace et al., 1999)。可以在水生和陆生食物网中找到自上而下控制的例子,顶层捕食者控制其猎物种类(初级捕食者和/或食草动物)的多度,进而控制食草动物的水平和基础物种的多度和多样性(Chase et al., 2000)。频率制约性捕食,即个体最丰富的物种比那些含较少个体的物种更有可能被吃掉,这也可以作为生态群落中的一个稳定机制,减少可能导致竞争排斥的猎物物种之间的适应性差异(Chesson, 2000;见 4.2.3 小节)。

营养级联描述的情况是,强大的捕食者-食饵效应改变了一个物种、功能群或营养水平的多度、生物量或生产力,改变了食物网中的多个环节(Pace et al., 1999)。虽然营养级联通常开始于食物网的最高营养级别(即顶层捕食者),但它们也可以开始于较低的级别。城镇化过程可以在很大程

图 4.2 一个简化的城市食物网：1. 腐生生物（真菌）；2. 自养生物（植物）；3. 食草动物或主要消费者（蚂蚁、鸽子、蛾子）；4. 初级捕食者（黑鸟、食虫蝙蝠）；5. 次级捕食者（猫和狐狸）

度上改变不同营养层次的物种多度，从而改变陆地和水生食物网的组成、结构和功能（Faeth et al.，2005）。因此，城镇化会破坏现有的营养级联，因为城镇化会减少顶层捕食者的数量，如大型哺乳类食肉动物，然后导致次级、中型捕食者（也称为中捕食者）增多（Ritchie，Johnson，2009），随之，初级食肉动物和食草动物的数量减少。在食物网的另一端，城市生境中水和营养物质输入的变化可能导致初级生产力迅速增加，并可能加强对食草动物种群的自下而上效应，而增加的食物资源可能会改变多个营养级的竞争和捕食性相互作用（Faeth et al.，2005；Ritchie，Johnson，2009）。

梯度范式（Whittaker，1967），认为环境（非生物）变化在空间上是有序的，当向特定方向上移动时，给定的环境变量通常有渐进的变化。环境变化的空间格局会影响种群和物种的空间分布以及群落的组成（Terborgh，1971；Austin，1987；

Buckley, Jetz, 2010; Hoverman et al., 2011)。环境梯度的例子包括温度梯度和降雨量梯度,而温度梯度和降雨量梯度又可能与海拔梯度、纬度梯度有关(Bennett et al., 1991; Weaver, 2000; Potapova, Charles, 2002; Hu et al., 2010)。McDonnell 和 Pickett(1990)首次提出了城乡梯度的概念,认为城市和乡村处于人类对地球及其生态系统影响梯度的相反两端。将"城市性"食物类似于其他环境梯度的梯度,而不是视为简单的二元分类(无论是否为城市),都代表了城市生态学的重要进步。城乡梯度通常是一个复杂的梯度,代表了许多从城市环境向乡村环境转移时随空间变化的环境变量(McDonnell, Pickett, 1990; Hahs, McDonnell, 2006)。例如,建筑密度和高度、不渗透表面的覆盖比例、各种大气污染物的浓度,以及道路的密度。城乡梯度也可以用温度梯度来表征,气温随着梯度从城市向乡村的一端移动,不断下降(见第 2 章; Howard, 1833; Bridgman et al., 1995; Torok et al., 2001)。另外,所有这些环境变量都与城市或城市区域的人口密度高度相关。

城乡梯度已作为一个框架,用于城市栖息地中广泛的生态群落实证研究。城乡梯度的早期概念趋于线性化,描述了高度城镇化的"核心"或中央商务区,周围是由内郊区、外郊区到城乡边缘而构成城镇化逐渐减弱的同心环(见 McDonnell et al., 1993; Luck, Wu, 2002)。因此,一些学者批评城乡梯度是对城市空间复杂性的过度简化(见 Alberti, 2008; Ramalho, Hobbs, 2012)。然而,城乡梯度不一定是线性的。相反,它们可以采取多种形式,具体取决于特定城市或地理区域中城镇化的强度和空间模式。与其他类型的环境梯度一样,城乡梯度也研究种群和物种的分布,影响分布的过程以及由此产生的群落组成的空间变化(见 Blair, 1996; Niemelä et al., 2002; Lawson et al., 2008; Stracey, Robinson, 2012; Hamer, Parris, 2013)。

McDonnell 和 Hahs(2008)综述研究 200 多项关于城乡梯度不同分类群,包括鸟类、昆虫、哺乳动物、鱼类、植物和真菌。正如其他环境梯度的观察,每个分类组内的不同物种或群落对城乡梯度的反应不同。例如,一项针对南非比勒陀利亚城乡梯度中鸟类群落的研究发现,半自然栖息地(位于梯度的乡村端)共有 65 种原种生物,而郊区栖息地和城市栖息地中分别有 45 种和 47 种(van Rensburg et al., 2009)。随着城镇化进程而消失的物种,往往是草原专性物种,像非洲野鹟(*Saxicola torquatus*)、棕颈歌百灵(*Mirafra africana*)、非洲鹌鹑雀(*Ortygospiza atricollis*)(图 4.3)。然而,半自然生境中缺少的其他本土物种,在郊区和城市地

4.2 选择:城市生态学中的生态位理论

图4.3 (a)南非比勒陀利亚;(b)棕颈歌百灵(*Mirafra africana*),一种草地专属鸟类,在城镇化后消失了;(c)白腹太阳鸟(*Cinnyris talatala*),在城市和郊区发现[摄影:(a)Petrus Potgieter;(b)(c)Derek Keats]

区出现了中等多度。该类型包括画眉(*Turdus smithi*)、金头缝叶莺(*Spermestes cucullatus*)和白腹太阳鸟(*Cinnyris talatala*)。非本土物种,例如岩鸽和家麻雀,在半自然栖息地基本不存在,但在城市栖息地却很丰富;这些物种,有时被称为亲人类物种或城市剥削者(见专栏7.1)。总之,当梯度从乡村向城市的另一端移动时,鸟类的总多度增加了(van Rensburg et al., 2009)。

这些发现与世界其他城市的发现相似;在相对丰富和可预测的食物、水的供应支持下,城乡梯度中城市末端的鸟类群落,往往具有较低的物种丰富度,但是个体密度较高(见 Marzluff,2001;Shochat,2004;Shochat et al., 2004, 2010;Chace, Walsh, 2006;图4.4)。在整个城乡梯度范围内观察到的其他常见的群落模式(McKinney, 2008),包括物种丰富度随着城镇化程度的提高而持续下降(例

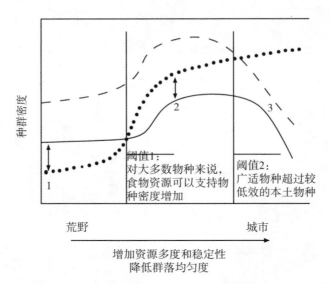

图4.4 在城市鸟类群落的常见模式示意图;沿城乡梯度的种群密度(实线和点线)和物种多样性(虚线段)的变化。1. 在野外,本土物种胜过入侵的亲人类物种(点线)。2. 在城镇化中等水平区域,两组物种仍然存在。通过利用新的、稳定和同质的食物资源,入侵物种的密度比本土物种增加得更快。本土物种通过空间划分而得以持续存在,利用残余斑块仍含有特殊的食物或其他资源,但长期来看可能会减少。3. 在高水平的城镇化进程中,共存机制崩溃,捕食压力减少,入侵物种成为主要的觅食者,而特有的本土物种则在当地灭绝(Shochat et al., 2010,图5b,经牛津大学出版社许可转载)

如，11项欧洲、北美洲、南美洲和澳大利亚的两栖动物研究）；在沿梯度的中间点，物种丰富度达到峰值，通常本土植被和郊区发展之间存在一个界面，导致栖息地多样性增加（例如，在欧洲和北美洲进行的6项研究）；在特定城镇化的阈值下，物种丰富度急剧下降（例如，澳大利亚墨尔本的河流大型无脊椎动物；Blair, Launer, 1997; Walsh et al., 2005, 2007; Sadler et al., 2006）。由此看来，分析的尺度可以极大地影响沿城乡梯度观察到的物种丰富度格局，在设计或研究时应考虑这一点（McKinney, 2008）。此外，沿城乡梯度生态群落的一些研究在此问题层面（即研究点在梯度上的位置）的研究样本不足，可能会限制其推理能力（Blair, 1996, 1999）。

4.2.2 栖息地模型

栖息地模型是一个概念或统计模型，其将一个响应变量（例如物种的出现概率或多度，生态群落的物种丰富度）与一个或多个环境变量相关联（Wintle et al., 2005）。栖息地模型可以帮助生态学家识别、确定重要的生态位关系，并使之规范化（Elith, Leathwick, 2009）。它们还可以通过估计某个物种或群落在景观的不同部分中出现的概率，来预测某个物种或群落适宜的栖息地位置，这是保护规划中的重要应用（Ferrier et al., 2002a, 2002b）。即使有可能对给定区域中每个位置的每种感兴趣的物种进行实地调查，这样做也非常耗时、昂贵。在规划不同土地用途的位置时（例如保护区、木材砍伐区、城市发展区），构建结构合理、可靠且防御性的栖息地模型，可以减少对综合野外数据采集的需求。

概念栖息地模型（称为生境适应性指数）可以用很少或不用现场数据来构建，而是依靠专家意见来确定感兴趣的物种或群落与影响其分布的环境变量之间的关系（Burgman et al., 2001）。然而，这些模型的质量在很大程度上取决于用于构建它们的专家意见的质量，并且缺乏独立的现场数据可能很难评估和维护它们（Wintle et al., 2005）。如果可以获得有关物种或群落的空间分布，以及一个或多个相关环境变量的足够多的数据，则生态学家可以构建统计栖息地模型。这些例子包括适用于非计划的、仅存在数据的模型（如 BIOCLIM, DOMAIN 和 MAXENT 等）(Busby, 1991; Carpenter et al., 1993; Phillips et al., 2006; Elith et al., 2006)。

广义线性模型(例如逻辑回归模型和泊松回归模型),通常需要跨空间中每个物种或群落是否存在的信息,即系统调查收集的数据(Wintle et al., 2005)。逻辑回归模型使用二进制、存在/缺失的数据来评估物种或群落的出现概率,而泊松回归模型(负二项式的特殊情况)可以使用统计数据来评估物种的多度和一个群落的物种丰富度,以此作为一个环境变量函数(Parris, 2001; Parris, 2006; Hamer, Parris, 2011)。多元关联(或排序)方法,如主成分分析和典范对应分析,将一个群落中多物种的存在/缺失或多度与环境变量相关联(Ter Braak, 1986, 1987; Ter Braak, Prentice, 1988)。部分排序方法(如部分冗余分析)计算由环境、空间和空间结构环境变量解释的群落组成变化量(Borcard et al., 1992; Cottenie, 2005; Peres-Neto, Kembel, 2015)。

栖息地模型已应用于研究城市生态学中的各种问题,包括景观尺度环境变量对日本野兔(*Lepus brachyurus*)沿城乡梯度出现概率的影响(Saito, Koike, 2009),评估南非夸祖鲁-纳塔尔省快速城镇化地区濒临灭绝的黑头矮变色龙(*Bradypodion melanocephalum*)的分布(Armstrong, 2009),以及物种水平特征对11个城市中超过8000种植物持续存在或局部灭绝概率的影响(Duncan et al., 2011;图4.5)。

泊松回归模型表明,澳大利亚墨尔本市区湿地中的树蛙(*Litoria* spp.)数量与捕食性鱼类的多度之间存在强烈的负相关关系(图4.6)。一项在肯尼亚基苏木和马林迪的研究表明,将栖息地模型应用于研究城市环境中人类的健康问题时,逻辑回归模型和泊松回归模型均被用于识别按蚊(*Anopheles*)(一种传播疟疾的蚊子)栖息地相关的环境变量(Jacob et al., 2005)。按蚊属蚊子是疟疾原虫的媒介,城市发展和扩张很可能会改变其幼虫的潜在栖息地的性质和分布(Jacob et al., 2005)。

栖息地模型也可用于识别城镇化景观中影响某些栖息地类型或生态群落持久性的环境变量。案例研究包括中国北京城市边缘的耕地,被转化为居住用地或工业用地的概率(Liu et al., 2013),以及由于澳大利亚墨尔本郊区的外来草入侵而使濒临灭绝的草地斑块消失或退化的概率(Williams et al., 2005)。在第一个案例中,2004年至2007年,研究区域的城市发展占用了超过5000hm^{2}的农田。与当前用于种植农作物或蔬菜的农田相比,休耕地更可能被转换为居住或工业用地

4.2 选择：城市生态学中的生态位理论

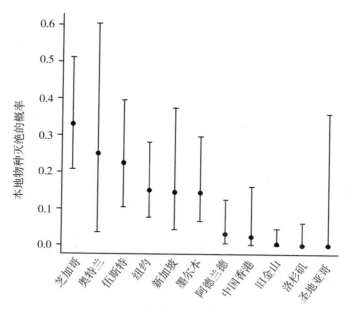

图 4.5 以 95% 置信区间评估每个分类性状在参考类中具有固定特征（草本、一年生、非克隆、非生物授粉、无特殊的扩散、固氮、C3 光合途径、无刺）的植物物种的局部灭绝概率（Duncan et al., 2011；图 3）

（Liu et al., 2013）。在第二个案例中，尽管研究区域的植被群落已濒临灭绝，但是在 1985 年至 2000 年，城市开发却摧毁了研究区域内 1670hm² 或占总面积 23% 的温带草地。在这两个案例中，与现有开发项目或主要道路的空间相邻性增加了城市土地用途转变的可能性。

4.2.3 生态组团和资源竞争模型

生态组团是一组同区域物种以相似的方式利用同一类环境资源，这些资源包括食物、庇护场所或筑巢场地（Root, 1967；Simberloff, Dayan, 1991）。将同区域物种划分为组团，而不是将生态群落中的所有物种都视为潜在的竞争者，是一种可以专注于研究具有特定功能关系的物种组的有效方法（Simberloff, Dayan, 1991；Williams, Hero, 1998；Luck et al., 2013；Gates et al., 2015）。生态学家已经将组团概念应用到城市环境中的群落以解决各种问题，包括哪些物种正在利

111

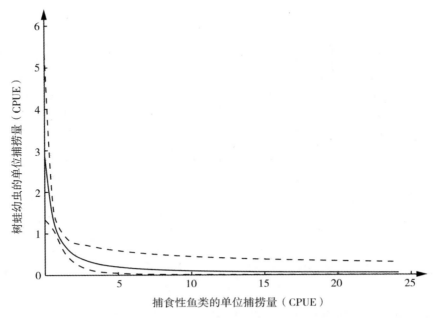

图4.6 澳大利亚墨尔本市区湿地中的树蛙(*Litoria* spp.)幼虫多度和捕食性鱼类的多度之间的关系,均以单位捕捞量(CPUE)表示。实线显示通过贝叶斯、零膨胀负二项式回归模型(ZINB)预测的关系,虚线显示95%置信区间(Hamer, Parris, 2013,图4a)

用/竞争城市中特定的资源;因城镇化进程改变或损失特定资源时,最有可能影响哪些物种;以及当一个或多个以前利用这些资源的物种在当地灭绝时,资源的可用性受到怎样的影响(Lim, Sodhi, 2004; Pauw, Louw, 2012; Hironaka, Koike, 2013; Luck et al., 2013; Huijbers et al., 2015)。

例如,新加坡一项对鸟类组团的研究发现,食虫性和食肉性鸟类(后一种主要以昆虫以外的动物为食)受城镇化不利影响时,食果鸟类则在人口密度低、大量种植果树的区域中大量繁殖(Lim, Sodhi, 2004)。随着城镇化水平的提高,在灌木丛中筑巢的和那些在树上挖洞筑巢的鸟类(如啄木鸟和翠鸟)会有所减少。在澳大利亚东南部的18个城镇中,适应在开阔和边缘栖息地觅食的食虫蝙蝠,比在混乱栖息地觅食的蝙蝠的数量多(Luck et al., 2013;图4.7)。适应杂乱区域的物种主要在茂密植被区觅食;该组团似乎对城镇化最为敏感。

4.2 选择：城市生态学中的生态位理论

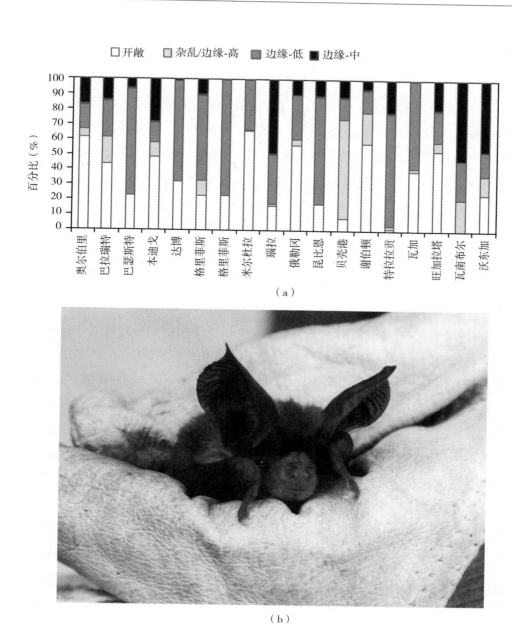

图4.7 （a）在澳大利亚东南部的18个城镇中，五个蝙蝠功能组团的百分率：适应开敞（开敞）；适应杂乱（杂乱）；适应边缘与低频回声定位（边缘-低）；适应边缘与中、高频回声定位（边缘-中或边缘-高）。此图中，已将适应杂乱和边缘-高的物种合并在一起（Luck et al.，2013；图3）。（b）长耳蝙蝠（*Nyctophilus gouldi*）（摄影：Tracy Lee）

当两个或两个以上具有基本生态位重叠的物种栖息在同一地理区域时，它们通常会争夺资源，例如：食物、水、庇护所或繁殖场地(迁移动物)；发芽部位、光照、土壤湿度或营养(植物)；栖息点、光照和食物(幼年期可以移动的无脊椎动物，例如一些海洋无脊椎动物)。这些竞争可能会对该地理区域每个物种的实际分布或现实生态位产生重要影响。如第4.2.1节所述，竞争排斥原理预测两个具有相同基础生态位的物种不能无限期共存，因为一个物种(优势竞争者)必须最终排除另一个物种(劣势竞争者)(Gause，1934；Hardin，1960)。因此，物种之间(或相似物种之间)的适应差异，会促进竞争性排斥(Chesson，2000)。

竞争排斥原理可以拓展到研究具有相似生态位的物种(考虑到在环境空间所有可能维度上难以确定两个生态位的等价性或其他性质)，以及在时间和空间上同时出现两个以上此类物种的情况。有趣的是，在自然环境中很少观察到这种竞争排斥，具有明显相同生态位的多个物种共存，在生态学上被认为是悖论(Hutchinson，1961；Bode et al.，2011)。栖息地可用性的时空变化、环境随机性、捕食者和寄生生物的介导作用，物种之间在距离或传播时间上的差异可能会导致具有相似生态位的多个物种共存(Shmida，Ellner，1984；Bode at al.，2011；Jiang et al.，2013)。Chesson(2000)提出了一个框架，将促进持续共存的机制分为两类：稳定机制，如生态性差异(也称为资源分配)，频率依赖性捕食，以及依赖时空环境变化的机制；平衡机制，可以减少物种之间的适应性差异。由于物种间的适应性差异，会导致竞争排斥，因此，它们与物种共存的稳定机制相反。

人们通常认为，在城市环境条件下的竞争优势是某些物种在城市中蓬勃发展而其他物种衰落或消失的原因(Shochat et al.，2004，2010；McKinney，2006)。但是，共存理论表明小规模的斑块(即小空间尺度上的环境异质性)和导致城市生境随时间变化的高频物理干扰，可能会帮助多个相似物种共存。假设共存理论适用于城市环境，那么主要的城市物种可能会比它们所替代的物种具有更大的适应性优势，这与城市环境条件的变化所产生的稳定机制相反。如前所述，Shochat等(2004，2010)认为，城市鸟类群落的形成是由于食物供应量的增加和捕食压力降低造成的(图4.4)。某些鸟类，如亚利桑那州凤凰城的印加鸠($Scardafellaina$)和哀鸽($Zenaida\ macroura$)，是城市栖息地中的高效觅食者，它们在这些栖息地中被捕食的风险要比在原始沙漠环境中低。这使印加鸠和哀鸽在城市环境中，相比其他本土食肉动物有更强的竞争优势。城市中某些捕食者的数量较少,也可能降低捕食对高多度物种的介导作用,这是另一种可以促进共存的稳定机制。然而,对相同资源的竞争应与城市环境中生物物理变化的影响区分开来,后者有利于某些具有基础生态位的物种而非其他物种(Marzluff,2005)。

4.3 生态漂移:模拟城市群落中的随机性

生态学中,中性模型强调生态漂移,它是由随机性或不确定性引起的。生态漂移在构建生态群落中的重要性一直是生态学争论的话题。Hubbell(2001)提出了生物多样性和生物地理学的统一中性理论,用以解释给定营养级的群落中的物种丰富度和相对多度的模式,其中,多个物种共享相似的生态位,并竞争相似的资源。如果认为这些物种在生态上是等效的(即每个物种的每个个体,都有可能繁殖、死亡、迁移或进化为新物种),那么随机性可能会对群落结构产生重大影响。统一中性理论的一个关键原则是,群落动态是一个"零-和"博弈——群落中的任何物种都不能在没有其他物种多度下降的情况下增加其多度(Hubbell,2001)。

在这种理论中,Hubbell(2001)使用一个中性词来描述单位资本生态等值性的假设,并将生态漂移定义为与种群统计的随机性基本相同(一个种群中出生和死亡的概率,见第2章和第3章)。尽管生态学家以前认识到这种随机性会影响物种的多度和物种的共存性,进而影响群落的构成(Chesson,Warner,1981;Adler et al.,2007),Hubbell的中性理论增加了人们对生态漂移的认识,将它作为群落生态学中的一个重要过程(Vellend,2010)。尽管在大多数生态群落中漂移不太可能是唯一的过程,但是它仍然可能是一个重要的过程,特别是在物种之间的功能差异不大的区域,即选择能力弱的区域(Vellend,2010)。生态漂移还可以与扩散、进化多样化相互作用,形成结构群落。例如,马来西亚的帕索森林保护区和巴拿马的巴罗科拉多岛的热带雨林中,$50hm^2$的固定样地上的树种相对多度,可以用一个包括生态漂移和扩散的中性模型来很好地描述(Hubbell,2001)。

但是,生态漂移对于塑造城市环境中的群落有多重要?显然,这在一定程度上取决于漂移相对于城镇化之前特定群落的选择、扩散、多样化的重要性。不过，这还取决于城镇化进程如何改变生态漂移与其他三个群落层次过程之间的平衡。城市基础设施(道路)导致的栖息地破碎化和孤岛化，会限制许多陆生动物以及不靠风传播种子的植物的扩散(Trom-bulak, Frissell, 2000; Develey, Stouffer, 2001; Desender et al., 2005; 见第 3 章)。这可能会增加生态漂移在塑造这些相对孤立的城市群落中的重要性。随着群落中个体总数的下降，生态漂移的重要性可能会增加(Vellend, 2010)。这类似于种群遗传学中的现象，随着种群的减少或出现瓶颈，遗传漂移加快了(Ellstrand, Elam, 1993)。某些类型的城市生态群落，可能比非城市群落遭受更强烈和更频繁的干扰(Ishitani et al., 2003)。反过来，这可能放大群落中不同物种偶然迁移或局部灭绝在塑造未来组成和相对多度中的重要性。另一方面，城镇化进程中诸如气候变化、水文状况或噪声、日照等，可能会增加对某些在新条件下可以耐受或繁衍的物种的选择(Raghu et al., 2000; Bonier et al., 2007)，而强有力的选择可能会抵消生态漂移的影响(Vellend, 2010)。

一项针对瑞士三个城市的蜘蛛、蜜蜂和鸟类群落的研究发现，群落的组成受环境变量的影响很大，如场地周围某一特定半径内的方位、太阳照射和木本植被比例，而不是空间变量(Sattler et al., 2010)。这说明塑造这些群落中物种的组成和相对多度方面，选择的作用比扩散的作用更强。然而，该研究结论解释群落结构的总体变化是相当有限的，从卢塞恩和卢加诺的蜜蜂群落为 4%，到卢加诺的鸟类群落为 30%，这九个群落的平均变异量为 12.5%。这意味着这些群落组成中 70%~96%的变化，无法通过考虑环境或空间变量来解释。这些群落组成的某些无法解释的变化可能是人类反复干扰城市环境而造成的生态漂移所致(Sattler et al., 2010)，这减少了选择和扩散对群落结构的影响。另一个解释是，该研究没有包括某些环境变量，但这些变量是每个群落组成的重要驱动因素。Sattler 等(2010)提出，他们研究的群落中几乎没有空间结构(因此扩散)，这是许多城市生态群落的典型特征，与其他在城市栖息地中对群落构成的研究相反，后者发现了群落结构中重要的空间组成部分(Parris, 2006; Chytrý et al., 2012)。显然，这需要做进一步的实验和理论研究，以确定漂移对塑造城市生态群落的重要性。

4.4 扩散：个体在空间中的移动

扩散是指个体从繁殖场所(种子和孢子)、幼虫或成虫，移动到另一场所。随着时间的推移，不同物种的个体到来或离开，扩散在构建生态群落中可以发挥关键作用。扩散对群落动态的影响取决于源群落(扩散个体的源地)和接收群落(到扩散的地方)的大小、位置和组成，以及每个群落的连通性或孤立性(Leibold et al.，2004；Vellend，2010)。反过来，连通性和孤立性取决于群落之间的地理距离以及个体成功穿越分隔地的概率(Leibold et al.，2004；Moritz et al.，2013)。在自然环境中，当地群落的组成受到区域群落组成的强烈影响，因为前者的物种必然是后者物种库的子集(图 4.8)。同样，区域群落构成也受到大陆或全球群落的强烈影响(Vellend，2010)。然而，在城市环境中人类可以将外来物种引入其区域群落中，从而增加其库源的有效规模——可能会非常大(图 4.8)。例如，这可以降低对区域种群库的依赖性，导致许多城市植物群落的物种丰富度提高(Hope et al.，2003；Knapp et al.，2008)。

群落生态学家提出了一些理论和模型，来解释扩散如何影响物种的共存以及群落组成的其他方面，包括物种丰富度和相对多度。其中一些理论只专注于扩散，但大多数理论将扩散与一个或多个其他群落层次过程(选择、生态漂移或进化多样化)相结合。例如，岛屿生物地理学的平衡理论考虑了漂移和扩散(MacArthur，Wilson，1963，1967)，生物多样性和生物地理学的统一中性理论考虑了生态漂移、物种形成和扩散(Hubbell，2001)，集合群落理论考虑了选择、生态漂移和扩散(Hanski，Gilpin，1991；Leibold et al.，2004)。Bode 等(2011)的差异扩散模型，致力于研究扩散和预测，如果同种物种所占据的栖息地斑块彼此之间的距离不规则，那么仅是种间扩散能力的差异，就可以使同种物种共存。

集合群落理论是集合种群理论的延伸(Levins，1969；见 3.2.1 节)，而后者又是由岛屿生物地理平衡理论发展而来的(MacArthur，Wilson，1963，1967)。一个集合群落可以定义为一组本地群落，它们占据着一组栖息地斑块，这些栖息地通过多种潜在相互作用的物种的扩散相联系(Leibold et al.，2004)。对集合群落的不同观点，不同程度地强调了扩散、选择和生态漂移的过程。当一个集合群落

第 4 章　群落响应城镇化

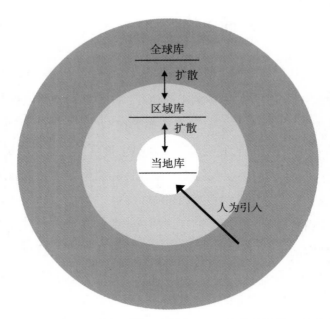

图 4.8　本地、区域和全球生态群落组成的示意图。通过物种的扩散，本地群落的组成与区域群落的组成相联系，而区域群落的组成与全球群落的组成相联系。但是，在城市环境中，本地物种库也通过人为引入区域群落中不存在的物种，而与全球物种库相关联

由被改变的或荒凉的景观所包围的永久性栖息地斑块组成时，随着局部灭绝的概率降低，栖息地内物种数量会随着斑块面积的增加而增加（物种-面积关系；Hanski，1994）。其次，由于与另一个斑块有机结合的可能性较低，因此孤立栖息地斑块的单位面积支持物种数量少于连接良好的斑块（物种-隔离关系；Hanski，1994）。如果环境条件随栖息地斑块的不同而变化，那么在不同栖息地斑块中可能有不同物种，从而扩散和选择都将影响本地群落的组成（物种-配置角度；Leibold et al.，2004）。在这种情况下，扩散仍然很重要，因为它可以使本地群落的组成发生变化，以适应本地环境条件的变化。集合群落理论的中性观点，强调生态漂移和随机扩散事件，所有物种在所有斑块中具有相似的适应性。

一项在澳大利亚墨尔本的研究评估了集合群落理论对城乡梯度上的 104 个池塘中蛙类群落的适用性（Parris，2006）。泊松回归模型研究发现，物种丰富度随斑块面积（池塘大小）的增大而增加，物种丰富度随斑块隔离度的增加而减少，

这是根据周围的道路覆盖物来衡量的。在所有其他变量保持不变的情况下，最大池塘的物种丰富度，预计比最小池塘高 2.8~5.5 倍，而最小孤立池塘的物种丰富度是最孤立池塘的 5.3~8.3 倍。环境条件也很重要，带有垂直围墙的池塘的物种丰富度，预计是有缓坡边缘池塘的物种丰富度的一半。在研究区内，垂直的墙壁使池塘不适合繁殖许多陆生青蛙，因为这些青蛙无法爬上垂直墙面；如果它们确实在有围墙的池塘中繁殖，它们的后代将被困在池塘中，在从蝌蚪变为幼蛙的过程中被淹死（Parris，2006）。南部的褐树蛙（*Litoria ewingii*）可以爬上垂直表面（图 4.9），并且是唯一经常栖息在有垂直墙壁的池塘中的物种。这些结果证明了集合群落理论在这种城市系统中的适用性，以及扩散和选择在构建池塘蛙群落中的重要性。关于扩散和选择影响城市湿地生态群落组成的进一步证据，是来自美国科罗拉多州的 201 个湿地两栖动物（Johnson et al.，2013）。所有这些分类种群都表明，城市区域的湿地物种丰富度，低于农业和草原地区，这是城市区域的环境变量和道路隔离池塘造成的。

图 4.9　澳大利亚墨尔本南部的褐树蛙（*Litoria ewingii*），仍然栖息在有垂直墙壁的池塘中（摄影：Kirsten M. Parris）

4.5 多样性：城市环境中新谱系的进化

多样性是指从现有谱系中进化出新谱系(包括新的基因型、变型、变种、亚种和种)。在细菌和古细菌(可以生活在极端环境中的细菌类型，如厌氧、耐高温)等微生物中，多样性通常是通过变异或水平基因转移产生的(Nemergut et al.，2013)。与大型生物相比，微生物可以非常迅速地进化，特别是在选择压力很大的情况下(Rainey, Travisano, 1998；Rensing et al.，2002；Nemergut et al.，2013)。例如，城市污水处理厂可以作为细菌对抗生素耐药性进化的研究热点(Auerbach et al.，2007；Novo et al.，2003；Rizzo et al.，2013)。抗生素残留、耐抗生素细菌和非耐药菌的结合，有利于废水中细菌个体之间对耐抗生素基因的选择和水平转移(Novo et al.，2013)。例如，在葡萄牙，经过处理的废水中的抗生素耐药性与原废水中四环素残留量和处理时高温呈正相关；不同菌群对处理过的废水中四环素和阿莫西林的耐药性在1%~60%之间变化(Novo et al.，2013)。在同一家工厂进行的一项早期研究发现，废水处理导致耐环丙沙星肠球菌的流行率增加了3倍，这可能是由于对环丙沙星具有高耐药性的肠球菌(*Enterococcus faecium*)的选择和水平基因转移的综合作用(da Silva et al.，2006)。

在大型生物中，一系列过程可能导致进化多样化，其中许多过程涉及一定程度的生殖隔离(Coyne, Orr, 2004；Etienne et al.，2007)。例如，当一个有性繁殖物种的种群在地理上被隔离，并且它们之间的基因流动大大减少时，就会发生异源多样性。随着时间的推移，种群之间的遗传差异会通过遗传漂移和/或选择(当适应当地条件)而增加，有时达到繁殖不相容的程度，因此被视为独立的物种。如果一个孤立的种群很小，那么建群者影响力的共同作用和快速遗传漂移可导致物种快速形成。这被称为边域性物种，最常见的情况是，只有一个特种的少数个体迁移到孤立的地方，如一个岛屿(Barton, Charlesworth, 1984；Seddon, Tobias, 2007)。一个物种的种群也可以被基因隔离，尽管共享同一地理区域(同域性物种形成)；这可以通过杂交和/或多倍体产生(Rieseberg et al.，2006；Mallet, 2007；Abbott et al.，2013)。在后一种情况下，细胞分裂过程中的错误，会导致种群中某些个体携带多组染色体，这些多倍体通常无法与亲本物种的二倍

4.5 多样性：城市环境中新谱系的进化

体个体成功繁殖。例如，自 1700 年以来，人类活动的干扰、杂交、多倍体的共同作用，已导致英国城市环境中至少三种新植物物种的进化（Thomas，2015）。

通过性选择过程也可以产生新的物种（Panhuis et al.，2001）。性别选是群体内个体差异交配成功的结果，这通常由雄性间的交配竞争，以及雌性对某些雄性特征的偏好共同驱动的。当配偶偏好和一种或多种第二性特征的改变，导致物种内的种群生殖隔离时，性选择会导致物种形成（Panhuis et al.，2001）。第二性特征是区分性别的特征，但不是繁殖所必需的。形态示例，包括许多雄鸟鲜艳的羽毛、雄狮的鬃毛和雄鹿的鹿角；行为示例，包括雄鸟的歌声和雄蛙的响亮叫声。第二性特征可能与雄性品质（如能量储备、身体状况、体型）相关。因此对于希望与之交配的雌性和与之竞争的雄性而言，第二性特征都是可靠的品质信号（Hoelzer，1989；Moller，Pomiankowski，1993）。

城镇化将促进异域性物种多样化，包括物种形成，即要满足以下两个条件，一个物种的城市种群就会与乡村种群隔离开来：城市种群中的遗传漂移和/或选择会使得种群适应当地条件，从而导致了城市种群和乡村种群之间的遗传分化。但是，在城镇化之后，很少有经过证实的异源物种形成的例子。正在进行的异源物种多样化的一个例子是美国亚利桑那州的家雀（*Carpodacus mexicanus*）。图森市的这个物种的城市种群的个体比附近萨瓜罗沙漠的乡村种群的个体有更长、更深和更强的喙（Badyaev et al.，2008）。城市地区的家雀通常会觅食喂食器中的葵花，葵花籽比沙漠地区大部分食用种子更大、更坚硬。引入这些新的食物来源，似乎对城市家雀的喙的发展和形态产生了很大的选择压力，从而导致城市和乡村家雀的喙的大小和咬合力存在差异。这些对城市食物来源的适应与种群间的遗传差异以及雄性求偶鸣声结构的差异有关——与沙漠鸟类相比，拥有较长、较深喙的城市鸟类鸣叫的频率较低（每秒鸣叫的次数较少）和频率范围较广（Badyaev et al.，2008）。

如果城市环境特征导致择偶偏好的改变，以及关键的第二性特征的相关变化，城市物种的进化多样化可以通过性选择而发生。在上述家雀的例子中，对城市食物源的形态适应（较长、较深的喙）和雄性求偶鸣叫声的变化相关，这是一种第二性特征，在雌性择偶中起着重要作用。声音上的这种差异可能会增强当地的形态适应能力，并减少城市和乡村家雀种群之间的基因流动（Badyaev et al.，

2008）。众所周知，许多其他鸟类在城市中和乡村栖息地的鸣叫方式不同，包括以较高的音调发声以减少来自城市低频噪声的干扰（频率偏移），鸣叫声更大（振幅偏移），发声短、鸣叫得更慢（每秒的音节更少，每次叫声的音节也更少）（Slabbekoorn，Peet，2003；Brumm，2004；Nemeth，Brumm，2009；Potvin et al.，2011；Potvin，Parris，2012；Parris，McCarthy，2013；见 3.2.5 小节）。然而，迄今为止，没有其他研究发现城市鸟鸣声的这些特征与城市、乡村种群之间的形态或遗传差异相关（Partecke et al.，2004；Leader et al.，2008；Potvin et al.，2013）。

相反，城市和乡村的鸟类之间的鸣叫声差异，可能是通过声音可塑性（改变鸣声特征以适应环境噪声的水平和频率等条件的能力）和/或文化进化（鸣声类型跨代的非随机传播）而引起的（Halfwerk，Slabbekoorn，2009；Luther，Baptista，2010）。像鸟鸣一样，雄蛙的鸣叫声在择偶和加强种间界限中起着重要作用（Blair，1964；Wells，1977）。在许多蛙类中，已知雄性蛙在城市噪声和道路交通噪声中的鸣叫频率（音调）较高（Parris et al.，2009；Hoskin，Goosem，2010；Cunnington，Fahrig，2010）。在此阶段，尚不清楚这些频率的变化是声音可塑性的结果，还是对高音调鸣叫声的性选择的早期迹象。

4.6 总结

本章考虑了城镇化对生态群落的影响，重点讨论了四个群落层次的过程：选择、生态漂移、扩散和进化多样化。尽管生态学家已经提出了各种各样的假设、理论和模型用于解释生态群落的结构，但是很少有人考虑到城市环境而提出这些假说、理论和模型。然而，本章证明许多现有的生态理论和模型（包括生态位理论、环境梯度、资源竞争模型、食物网、营养级联和集合群落理论），确实适用于城市环境，可以增强我们对城镇化对生态群落的影响的认识。其他群落理论也可能适用于这些环境，但是有待研究。城镇化通过改变城市环境条件来改变生态群落，这反过来又有利于某些物种。群落中物种特性和相对多度的最终变化，会改变不同营养级（包括自养生物、食草动物和捕食者）物种之间的相互作用，从而导致某些物种的相对多度的进一步变化和/或局部灭绝。道路和其他城市基础设施造成栖息地碎片化和隔离，可能会限制个体从一个栖息地斑块向另外一个栖息地斑块扩散，从而对一系列生物分类种群产生重要影响。扩散程度降低，会增加随机过程(如统计随机性)对群落结构的影响。在城市栖息地中也有进化多样化的证据，即城市中某物种种群适应了当地条件或与乡村种群繁殖隔离。在下一章，我们将考虑和下一个层次的生态组织，研究整个生态系统对城镇化的响应。

思考题

1. 描述本章讨论的四个群落层次过程。你是否认为其中一些在城市环境中比其他更为重要？
2. 举例说明生态位理论在城市生态群落研究中的应用。
3. 画出你所了解的城市的食物网。
4. 城镇化如何影响扩散？整个城市景观的分散变化将如何影响生态群落的结构？
5. 城镇化进程是否增加了生态漂移的重要性？
6. 城镇化如何促进进化多样化，包括在城市中的物种形成？
7. 举例说明三种适用于城市环境中的群落生态学理论。

本章参考文献

Abbott R, Albach D, Ansell S, et al. (2013) Hybridization and speciation. *Journal of Evolutionary Biology*, 26, 229-246.

Adler P B, Hille RisLambers J, Levine J M. (2007) A niche for neutrality. *Ecology Letters*, 10, 95-104.

Alberti M. (2008) *Advances in Urban Ecology: Integrating Human and Ecological Processes in Urban Ecosystems*. Springer, New York.

Armstrong A J. (2009) Distribution and conservation of the coastal population of the black-headed dwarf chameleon *Bradypodion melanocephalum* in KwaZulu-Natal. *African Journal of Herpetology*, 58, 85-97.

Auerbach E A, Seyfried E E, McMahon K D. (2007) Tetracycline resistance genes in activated sludge wastewater treatment plants. *Water Research*, 41, 1143-1151.

Austin M P. (1987) Models for the analysis of species response to environmental gradients. *Vegetatio*, 69, 35-45.

Badyaev A V, Young R L, Oh K P, et al. (2008) Evolution on a local scale: developmental, functional, genetic bases of divergence in bill form and associated changes in song structure between adjacent habitats. *Evolution*, 62, 1951-1964.

Barton N H, Charlesworth B. (1984) Genetic revolutions, founder effects, speciation. *Annual Review of Ecology and Systematics*, 15, 133-164.

Begossi A. (1996) Use of ecological methods in ethnobotany: diversity indices. *Economic Botany*, 50, 280-289.

Bennett A F, Lumsden L F, Alexander J S A, et al. (1991) Habitat use by arboreal mammals along an environmental gradient in north-eastern Victoria. *Wildlife Research*, 18, 125-146.

Blair R B. (1996) Birds and butterflies along an urban gradient: surrogate taxa for assessing biodiversity? *Ecological Applications*, 9, 164-170.

Blair R B. (1999) Land use and avian species diversity along an urban gradient. *Ecological Applications*, 6, 506-519.

Blair R B, Launer A E. (1997) Butterfly diversity and human land use: species assemblages along an urban gradient. *Biological Conservation*, 80, 113-125.

Blair W F. (1964) Isolating mechanisms and interspecies interactions in anuran amphibians. *Quarterly Review of Biology*, 39, 333-344.

Bode M, Bode L, Armsworth P R. (2011) Different dispersal abilities allow reef fish to coexist. *Proceedings of the National Academy of Sciences of the United States of America*, 108, 16317-16321.

Bolnick D I, Amarasekare P, Araujo M S, et al. (2011) Why intraspecific trait variation matters in community ecology. *Trends in Ecology and Evolution*, 26, 183-192.

Bonier F, Martin P R, Wingfield J C. (2007) Urban birds have broader environmental tolerance. *Biology Letters*, 3, 670-673.

Borcard D, Legendre P, Drapeau P. (1992) Partialling out the spatial component of ecological variation. *Ecology*, 73, 1045.

Bridgman H, Warner R, Dodson J. (1995) *Urban Biophysical Environments*. Oxford University Press, Melbourne.

Brumm H. (2004) The impact of environmental noise on song amplitude in a territorial bird. *Journal of Animal Ecology*, 73, 434-440.

Buckley L B, Jetz W. (2010) Lizard community structure along environmental gradients. *Journal of Animal Ecology*, 79, 358-365.

Burgman M A, Breininger D R, Duncan B W, et al. (2001) Setting reliability bounds on habitat suitability indices. *Ecological Applications*, 11, 70-78.

Burke C, Steinberg P, Rusche D, et al. (2011) Bacterial community assembly based on functional genes rather than species. *Proceedings of the National Academy of Sciences*, 108, 14288-14293.

Busby J R. (1991) A bioclimate analysis and prediction system. In *Nature Conservation: Cost Effective Biological Surveys and Data Analysis*. Margules C R, Austin M P eds. CSIRO, Canberra: 64-68.

Carpenter G, Gillison A N, Winter J. (1993) A flexible modelling procedure for mapping potential distributions of plants and animals. *Biodiversity and Conservation*, 2, 667-680.

Chace J F, Walsh J J. (2006) Urban effects on native avifauna: a review. *Landscape and Urban Planning*, 74, 46-69.

Chase J M, Liebold M A, Downing A L, et al. (2000) The effects of productivity, herbivory, plant species turnover in grassland food webs. *Ecology*, 81, 2485-2497.

Chesson P L. (2000) Mechanisms of maintenance of species diversity. *Annual Review of Ecology and Systematics*, 31, 343-366.

Chesson P L, Warner R R. (1981) Environmental variability promotes coexistence in lottery competitive systems. *The American Naturalist*, 117, 923-943.

Chytry M, Lososova Z, Horsak M, et al. (2012) Dispersal limitation is stronger in communities of microorganisms across Central European cities. *Journal of Biogeography*, 39, 1101-1111.

Clements F E. (1916) *Plant Succession: An Analysis of the Development of Vegetation*. Carnegie Institute of Washington, Washington DC.

Clements F E. (1936) Nature and structure of the climax. *Journal of Ecology*, 24, 252-284.

Cottenie K. (2005) Integrating environmental and spatial processes in ecological community dynamics. *Ecology Letters*, 8, 1175-1182.

Coyne J A, Orr H A. (2004) Speciation, Sinauer Associates, Sunderland, MA.

Cunnington G M, Fahrig L. (2010) Plasticity in the vocalizations of anurans in response to traffic. *Acta Oecologica*, 36, 463-470.

da Silva M F, Tiago I, Verissimo A, et al. (2006) Antibiotic resistance of enterococci and related bacteria in an urban wastewater treatment plant. *FEMS Microbiology Ecology*, 55, 322-329.

Desender K, Small E, Gaublomme E, et al. (2005) Rural-urban gradients and the population genetic structure of woodland ground beetles. *Conservation Genetics*, 6, 51-62.

Develey P F, Stouffer P C. (2001) Effects of roads on movements by understory birds in mixed-species flocks in Central Amazonian Brazil. *Conservation Biology*, 15, 1416-1422.

Duncan R P, Clemants S E, Corlett R T, et al. (2011) Plant traits and extinction in urban areas: a meta-analysis of 11 cities. *Global Ecology and Biogeography*, 20, 509-519.

Elith J, Graham C H, Anderson R P, et al. (2006) Novel methods improve prediction of species distributions from occurrence data. *Ecography*, 29, 129-151.

Elith J, Leathwick J R. (2009) Species distribution models: ecological explanation and prediction across space and time. *Annual Review of Ecology, Evolution, Systematics*, 40, 677-697.

Ellstrand N C, Elam D R. (1993) Population genetic consequences of small population size: implications for plant conservation. *Annual Review of Ecology and Systematics*, 24, 217-242.

Etienne R S, Emile M, Apol F, et al. (2007) Modes of speciation and the neutral theory of biodiversity. *Oikos*, 116, 241-258.

Faeth S H, Warren P S, Shochat E, et al. (2005) Trophic dynamics in urban communities. *BioScience*, 55, 399-407.

Ferrier S, Drielsma M, Manion G, et al. (2002) Extended statistical approached to modelling spatial pattern in biodiversity in northeast New South Wales. II. Community-level modelling. *Biodiversity and Conservation*, 11, 2309-2338.

Ferrier S, Watson G, Pearce J, et al. (2002) Extended statistical approached to modelling spatial patterns in biodiversity in northeast New South Wales. I. Species-level modelling. *Biodiversity and Conservation*, 11, 2275-2307.

Gates K K, Vaughn C C, Julian J P. (2015) Developing environmental flow recommendations for freshwater mussels using the biological traits of species guilds. *Freshwater Biology*, 60, 620-635.

Gause G F. (1934) *The Struggle for Existence*. Williams & Wilkins, Baltimore.

Gerhardt H C, Huber F. (2002) *Acoustic Communication in Insects and Anurans: Common Problems and Diverse Solutions*. University of Chicago Press, Chicago.

Gleason H A. (1917) The structure and development of the plant association. *Bulletin of the Torrey Botanical Club*, 44, 463-481.

Gleason H A. (1926) The individualistic concept of the plant association. *Bulletin of the Torrey Botanical Club*, 53, 1-20.

Grinnell J. (1917) The niche-relationships of the California Thrasher. *The Auk*, 34, 427-433.

Hahs A K, McDonnell M J. (2006) Selecting independent measures to quantify Melbourne's urban-rural gradient. *Landscape and Urban Planning*, 78, 435-448.

Halfwerk W, Slabbekoorn H. (2009) A behavioural mechanism explaining noise-dependent frequency use in urban birdsong. *Animal Behaviour*, 78, 1301-1307.

Hamer A J, Parris K M. (2013) Predation modifies larval amphibian communities in urban wetlands. *Wetlands*, 33, 641-652.

Hamer A J, Parris K M. (2011) Local and landscape determinants of amphibian communities in urban ponds. *Ecological Applications*, 21, 378-390.

Hanski I. (1994) A practical model of metapopulation dynamics. *Journal of Animal Ecology*, 63, 151-162.

Hanski I. and Gilpin M. (1991) Metapopulation dynamics: brief history and conceptual domain. In *Metapopulation Dynamics*. Gilpin M E, Hanski I eds. Academic Press, London: 3-16.

Hardin G. (1960) The competitive exclusion principle. *Science*, 131, 1292-1297.

Hatcher M J, Dick J T A, Dunn A M. (2006) How parasites affect interactions between competitors and predators. *Ecology Letters*, 9, 1253-1271.

Hironaka Y, Koike F. (2013) Guild structure in the food web of grassland arthropod communities along an urban-rural landscape gradient. *Ecoscience*, 20, 148-160.

Hoelzer G A. (1989) The good parent process of sexual selection. *Animal Behaviour*, 38, 1067-1078.

Hope D, Gries C, Zhu W, et al. (2003) Socioeconomics drive urban plant diversity. *Proceedings of the National Academy of Sciences of the United States of America*, 100, 8788-8792.

Hoskin C J, Goosem M W. (2010) Road impacts on abundance, call traits, body size of rainforest frogs in northeast Australia. *Ecology and Society* 15, 15. http://www.ecologyandsociety.org/vol15/iss3/art15.

Hoverman J T, Davis C J, Werner E E, et al. (2011) Environmental gradients and the structure of freshwater snail communities. *Ecography*, 34, 1049-1058.

Howard L. (1833) *Climate of London Deduced from Meteorological Observations*. Harvey and Darton, London.

Hu L, Li M, Li Z. (2010) Geographical and environmental gradients of lianas and vines in China. *Global Ecology and Biogeography*, 19, 554-561.

Hubbell S P. (2001) *The Unified Neutral Theory of Biodiversity and Biogeography*. Princeton University Press, Princeton.

Huijbers C M, Schlacher T A, Schoeman D S, et al. (2015) Limited functional redundancy in vertebrate scavenger guilds fails to compensate for the loss of raptors from urbanized sandy beaches. *Diversity and Distribution*, 21, 55-63.

Hutchinson G E. (1957) Concluding remarks. *Cold Spring Harbor Symposium on Quantitative Biology*, 22, 415-427.

Hutchinson G E. (1961) The paradox of the plankton. *The American Naturalist*, 95, 137-145.

Ishitani M, Kotze J, Niemela J. (2003) Changes in carabid beetle assemblages across an urban-rural gradient in Japan. *Ecography*, 26, 481-489.

Jacob B G, Arheart K L, Griffith D A, et al. (2005) Evaluation of environmental data for identification of *Anopheles* (Diptera: Culicidae) aquatic larval habitats in Kisumu and Malindi, Kenya. *Entomology Society of America*, 42, 751-755.

Jiang T, Lu G, Sun K, et al. (2013) Coexistence of *Rhinolophus affinis* and *Rhinolophus pearsoni* revisited. *Acta Theriologica*, 58, 47-53.

Johnson P T J, Hoverman J T, McKenzie V J, et al. (2013) Urbanization and wetland communities: applying metacommunity theory to understanding the local and landscape effects. *Journal of Applied Ecology*, 50, 34-42.

Knapp S, Kuhn I, Schweiger O, et al. (2008) Challenging urban species diversity: contrasting phylogenetic patterns across plant functional groups in Germany. *Ecological Letters*, 11, 1054-1064.

Konstantinidis K T, Ramette A, Tiedje J M. (2006) The bacterial species definition in the genomic era. *Philosophical Transactions of the Royal Society B*, 361, 1929-1940.

Lawson D M, Lamar C K, Schwartz M W. (2008) Quantifying plant population persistence in human-dominated landscapes. *Conservation Biology*, 22, 922-928.

Lawton J. (1999) Are there general laws in ecology? *Oikos*, 84, 177.

Leader N, Geffen E, Mokady O, et al. (2008) Song dialects do not restrict gene flow in an urban population of the orange-tufted sunbird, *Nectarinia osea*. *Behavioral Ecology and Sociobiology*, 62, 1299-1305.

Leibold M A, Holyoak M, Mouquet N, et al. (2004) The metacommunity concept: a frame-work for multi-scale community ecology. *Ecology Letters*, 7, 601-613.

Levins R. (1969) Some demographic and genetic consequences of environmental heterogeneity for biological control. *Bulletin of the Entomological Society of America*, 15, 237-240.

Lim H C, Sodhi N S. (2004) Responses of avian guilds to urbanization in a tropical city. *Landscape and Urban Planning*, 66, 199-215.

Liu X, Zhang W, Li H, et al. (2013) Modeling patch characteristics of farmland loss for site assessment in urban fringe of Beijing, China. *Chinese Geographical Science*, 23, 365-377.

Luck G W, Smallbone L, Threlfall C G, et al. (2013) Patterns in bat functional guilds across multiple urban centres in south-eastern Australia. *Landscape Ecology*, 28, 455-469.

Luck M, Wu J. (2002) A gradient analysis of urban landscape pattern: a case study from the Phoenix metropolitan region, Arizona, USA. *Landscape Ecology*, 17, 327-339.

Luther D, Baptista L. (2010) Urban noise and the cultural evolution of bird songs. *Proceedings of the Royal Society B*, 277, 469-473.

MacArthur R H. (1960) On the relative abundance of species. *The American Naturalist*, 94, 25-36.

MacArthur R H, Wilson E O. (1963) An equilibrium theory of insular zoogeography. *Evolution*, 17, 373-387.

MacArthur R H, Wilson E O. (1967) *The Theory of Island Biogeography*. Princeton University Press, Princeton.

Malle J. (2007) Hybrid speciation. *Nature*, 446, 279-283.

Marzluff J M. (2005) Island biogeography for an urbanizing world: how extinction and colonization may determine biological diversity in human-dominated landscapes. *Urban Ecosystems*, 8, 157-177.

Marzluff J M. (2001) Worldwide urbanization and its effects on birds. In *Avian Ecology and Conservation in an Urbanizing World*. Marzluff J M, Bowman R, Donnelly R eds. Kluwer, Boston, 19-38.

McDonnell M J, Hahs A K. (2008) The use of gradient analysis studies in advancing our understanding of the ecology of urbanizing landscapes: current status and future directions. *Landscape Ecology*, 23, 1143-1155.

McDonnell M J, Pickett S T A. (1990) Ecosystem structure and function along urban-rural gradients: an unexploited opportunity for ecology. *Ecology*, 71, 1232-1237.

McDonnell M J, Picket S T A, Groffman P M, et al. (1993) Ecosystem processes along an urban-to-rural gradient. *Urban Ecosystems*, 1, 21-36.

McKinney M L. (2008) Effects of urbanization on species richness: a review of plants and animals. *Urban Ecosystems*, 11, 161-176.

McKinney M L. (2006) Urbanization as a major cause of biotic homogenization. *Biological Conservation*, 127, 247-260.

Moller A P, Pomiankowski A. (1993) Why have birds got multiple sexual ornaments? *Behavioral Ecology and Sociobiology*, 32, 167-176.

Moritz C, Meynard C N, Devictor V, et al. (2013) Disentangling the role of connectivity, environmental filtering, spatial structure on metacommunity dynamics. *Oikos*, 122, 1-10.

Nemergut D R, Schmidt S K, Fukami T, et al. (2013) Patterns and processes of microbial community assembly. *Microbiology and Molecular Biology Reviews*, 77, 342-356.

Nemeth E, Brumm H. (2009) Birds and anthropogenic noise: are urban songs adaptive? *The American Naturalist*, 176, 465-475.

Niemela J, Kotze D J, Venn S, et al. (2002) Carabid beetle assemblages(Coleoptera, carabidae) across urban-rural gradients: an international comparison. *Landscape Ecology*, 17, 387-401.

Novo A, Andre S, Viana P, et al. (2013) Antibiotic resistance, antimicrobial residues and bacterial community composition in urban wastewater. *Water Research*, 47, 1875-1887.

Pace M L, Cole J J, Carpenter S R, et al. (1999) Trophic cascades revealed in diverse ecosystems. *Trends in Ecology and Evolution*, 14, 483-488.

Palmer M W. (1994) Variation in species richness: towards a unification of hypotheses. *Folia Geobotanica and Phytotaxonomica*, 29, 511-530.

Panhuis T M, Butlin R, Zuk M, et al. (2001) Sexual selection and speciation. *Trends in Ecology and Evolution*, 16, 364-371.

Parris K M. (2001) Distribution, habitat requirements, conservation of the cascade treefrog (*Litoria pearsoniana*, Anura: Hylidae). *Biological Conservation*, 99, 285-292.

Parris K M. (2006) Urban amphibian assemblages as metacommunities. *Journal of Animal Ecology*, 75, 757-746.

Parris K M, McCarthy M A. (2013) Predicting the effects of urban noise on the active space of avian vocal signals. *The American Naturalist*, 182, 452-464.

Parris K M, Velik-Lord M, North J M A. (2009) Frogs call at a higher pitch in traffic noise. *Ecology and Society* 14, 25. http://www.ecologyandsociety.org/vol14/iss1/art25/.

Partecke J, Van't Hof T, Gwinner E. (2004) Differences in the timing of reproduction between urban and forest European blackbirds (*Turdus merula*): result of phenotypic flexibility or genetic differences? *Proceedings of the Royal Society B*, 271, 1995-2001.

Pauw A, Louw K. (2012) Urbanization drives a reduction in functional diversity in a guild of nectar-feeding birds. *Ecology and Society* 17, 27. http://www.ecologyandsociety.org/vol17/iss2/art27/.

Peres-Neto P R, Kembel S W. (2015) Phylogenetic gradient analysis: environmental drivers of phylogenetic variation across ecological communities. *Plant Ecology*, 216, 709-724.

Phillips S J, Anderson R P, Schapire R E. (2006) Maximum entropy modelling of species geographic distribution. *Ecological Modelling*, 190, 231-259.

Pimm S L, Lawton J H, Cohen J E. (1991) Food web patterns and their consequences. *Nature*, 350, 669-674.

Polis G A, Myers C A, Holt R D. (1989) The ecology and evolution of intraguild predation: potential competitors that eat each other. *Annual Review of Ecology and*

Systematics, 20, 297-330.

Potapova M G, Charles D F. (2002) Benthic diatoms in USA rivers: distribution along spatial and environmental gradients. *Journal of Biogeography*, 29, 167-187.

Potvin D A, Parris K M. (2012) Song convergence in multiple urban populations of silvereyes (*Zosterops lateralis*). *Ecology and Evolution*, 2, 1977-1984.

Potvin D A, Parris K M, Mulder R A. (2011) Geographically pervasive effects of urban noise on frequency and syllable rate of songs and calls in silvereyes (*Zosterops lateralis*). *Proceedings of the Royal Society B*, 278, 2464-2469.

Potvin D A, Parris K M, Mulder R A. (2013) Limited genetic differentiation between acoustically divergent populations of urban and rural silvereyes (*Zosterops lateralis*). *Evolutionary Ecology*, 27, 381-391.

Preston F W. (1948) The commonness, rarity, of species. *Ecology*, 29, 254-283.

Raghu S, Clarke A R, Drew R A I, et al. (2000) Impact of habitat modification on the distribution and abundance of fruit flies (*Diptera*: *Tephritidae*) in southeast Queensland. *Population Ecology*, 42, 153-160.

Rainey P B, Travisano M. (1998) Adaptive radiation in a heterogeneous environment. *Nature*, 394, 69-72.

Ramalho C E, Hobbs R J. (2012) Time for a change: dynamic urban ecology. *Trends in Ecology and Evolution*, 27, 179-188.

Rensing C, Newby D T, Pepper I L. (2002) The role of selective pressure and selfish DNA in horizontal gene transfer and soil microbial community adaptation. *Soil Biology and Biochemistry*, 34, 285-296.

Rieseberg L H, Wood T E, Baack E J. (2006) The nature of plant species. *Nature*, 440, 524-527.

Ritchie E G, Johnson C N. (2009) Predator interactions, mesopredator release, biodiversity conservation. *Ecology Letters*, 12, 982-998.

Rizzo L, Manaia C, Merlin C, et al. (2013) Urban wastewater treatment plants as hotspots for antibiotic resistant bacteria and genes spread into the environment: A review. *Science of the Total Environment*, 447, 345-360.

Root R B. (1967) The niche exploitation pattern of the blue-gray gnatcatcher. *Ecological Monographs*, 37, 317-350.

Rosindell J, Hubbell S P, Etienne R S. (2011) The unified neutral theory of biodiversity and biogeography at age ten. *Trends in Ecology and Evolution*, 26, 340-348.

Sadler J P, Small E C, Fiszpan H, et al. (2006) Investigating environmental variation and landscape characteristics of an urban-rural gradient using woodland carabid assemblages. *Journal of Biogeography*, 33, 1126-1138.

Saito M, Koike F. (2009) The importance of past and present landscape for Japanese hares *Lepus brachyurus* along a rural-urban gradient. *Acta Theriologica*, 54, 363-370.

Sattler T, Borcard D, Arlettaz R, et al. (2010) Spider, bee, bird communities in cities are shaped by environmental control and high stochasticity. *Ecology*, 91, 3343-3353.

Seddon N, Tobias J A. (2007) Song divergence at the edge of Amazonia: an empirical test of the peripatric speciation model. *Biological Journal of the Linnean Society*, 90, 173-188.

Shmida A, Ellner S. (1984) Coexistence of plant species with similar niches. *Vegetatio*, 58, 29-55.

Shochat E. (2004) Credit or debit? Resource input changes population dynamics of city-slicker birds. *Oikos*, 106, 622-626.

Shochat E, Lerman S B, Anderies J M, et al. (2010) Invasion, competition, biodiversity loss in urban ecosystems. *BioScience*, 60, 199-208.

Shochat E, Lerman S, Katti M, et al. (2004) Linking optimal foraging behavior to bird community structure in an urban-desert landscape: Field experiments with artificial food patches. *American Naturalist*, 164, 232-243.

Simberloff D. (2004) Community ecology: is it time to move on? *The American Naturalist*, 163, 787-799.

Simberloff D, Dayan T. (1991) The guild concept and the structure of ecological communities. *Annual Review of Ecology and Systematics*, 22, 115-143.

Slabbekoorn H, Peet M. (2003) Birds sing at a higher pitch in urban noise. *Nature*, 424, 267.

Spellerberg I F, Fedor P J. (2003) A tribute to Claude Shannon (1916-2001) and a plea for more rigorous use of species richness, species diversity, the "Shannon-Wiener" Index. *Global Ecology & Biogeography*, 12, 177-179.

Stracey C M, Robinson S K. (2012) Are urban habitats ecological traps for a native song-bird? Season-long productivity, apparent survival, site fidelity in urban and rural habitats. *Journal of Avian Biology*, 43, 50-60.

Ter Braak C J F. (1986) Canonical correspondence analysis: a new eigenvector technique for multivariate direct gradient analysis. *Ecology*, 67, 1167-1179.

Ter Braak C J F. (1987) The analysis of vegetation-environment relationship by canonical correspondence analysis. *Vegetatio*, 69, 69-77.

Ter Braak C J F, Prentice I C. (1988) A theory of gradient analysis. *Advances in Ecological Research*, 18, 271-317.

Terborgh J. (1971) Distribution on environmental gradients: theory and a preliminary interpretation of distributional patterns in the avifauna of the Cordillera Vilcabamba, Peru. *Ecology*, 52, 23-40.

Thomas C D. (2015) Rapid acceleration of plant speciation during the Anthropocence. *Trends in Ecology and Evolution*. 10.1016/j.tree.2015.05.009.

Thompson R M, Brose U, Dunne J A, et al. (2012) Food webs: reconciling the structure and function of biodiversity. *Trends in Ecology and Evolution*, 27, 689-697.

Torok S J, Morris C J G, Skinner C, et al. (2001) Urban heat island features of southeast Australian towns. *Australian Meteorological Magazine*, 50, 1-13.

Trombulak S C, Frissell C A. (2000) Review of ecological effects of roads on terrestrial and aquatic communities. *Conservation Biology*, 14, 18-30.

Van Rensburg B J, Peacock D S, Robertson M P. (2009) Biotic homogenization and alien bird species along an urban gradient in South Africa. *Landscape and Urban Planning*, 92, 233-241.

Vellend M. (2010) Conceptual synthesis in community ecology. *The Quarterly Review of Biology*, 85, 183-206.

Violle C, Enquist B J, McGill B J, et al. (2012) The return of the variance: intraspecific variability in community ecology. *Trends in Ecology and Evolution*, 27, 244-252.

Walsh C J, Fletcher T D, Ladson A R. (2005) Stream restoration in urban catchments through redesigning stormwater systems: looking to the catchment to save the stream. *Journal of the North American Benthological Society*, 24, 690-705.

Walsh C J, Waller K A, Gehling J, et al. (2007) Riverine invertebrate assemblages are degraded more by catchment urbanization than by riparian deforestation. *Freshwater Biology*, 52, 574-587.

Weaver P L. (2000) Environmental gradients affect forest structure in Puerto Rico's Luquilla Mountains. *Interciencia*, 25, 254-259.

Wells K D. (1977) The social behaviour of anuran amphibians. *Animal Behaviour*, 25, 666-693.

Whittaker R H. (1967) Gradient analysis of vegetation. *Biological Reviews*, 42, 207-264.

Williams N S G, McDonnell M J, Seager E J. (2005) Factors influencing the loss of an endangered ecosystem in an urbanising landscape: a case study of native grasslands from Melbourne, Australia. *Landscape and Urban Planning*, 71, 35-49.

Williams S E, Hero J-M. (1998) Rainforest frogs of the Australian wet tropics: guild classification and the ecological similarity of declining species. *Proceedings of the Royal Society B*, 265, 597-602.

Wintle B A, Elith J, Potts J M. (2005) Fauna habitat modelling and mapping: a review and case study in the Lower Hunter Central Coast region of NSW. *Austral Ecology*, 30, 719-738.

第5章 生态系统响应城镇化

5.1 概述

种群生态学可以通过出生、死亡、迁入和迁出这四个种群过程来理解;它关注的是群体中的个体。同样,可以通过选择、扩散、生态漂移和进化多样化的群落层次过程来理解群落生态学;它关注的是群落内的物种。但是,我们如何表征整个生态系统的生态学?首先,我们需要理清"生态系统"这个词的含义。"系统"是一组零件,它们通过一个或多个过程相互作用,具有从这些相互作用中出现新的性能。结果是系统作为一个整体大于各个部分的总和(Odum,1983)。Tansley(1935)将"生态系统"(ecological system)定义为:某一特定地点或地理区域中生物和非生物因素的集合,以及它们之间的相互作用网络。因此,生态系统具有四个主要的要素:生命有机体,包括动物、植物、微生物及人类;非生物元素,包括空气、水、光、热和矿物质(岩石、土壤、沙子、灰尘、碳和其他营养物质);生物之间、生物和非生物之间的相互作用;以及所有这些要素都存在的物理空间(Tansley,1935;Keith et al.,2013)。当生态系统的两个和多个组成部分相互作用时,就发生生物、化学和物理过程(Odum,1983);这些例子包括光合作用、固碳、蒸散、呼吸、捕食、固氮、分解、沉淀、日照、侵蚀,岩石和矿物的风化,以及土壤对水分和养分的吸收(见专栏5.1)。

生态系统生态学关注整个生态系统,对系统内外系统外的物质和能量流动进行测定和模拟(Loreau,2010)。像生态群落一样,生态系统的研究范围可以从局部到全球,取决于研究者感兴趣的问题和事物(Tansley,1935;Odum,1983;Willis,1997;Loreau,2010)。例如,让我们考虑流域或水文流域(集水区)形成的生态系统。流域历来是生态系统生态学中一个重要的分析尺度(Likens, 1992; Groffman et al., 2004),它足够大,可以观察系统内的能量、水和营养元素的循环;但是,又足够小,可以与环境管理相关。一个流域具有明确的物理边界(分水岭),这些边界定义了它的地理范围。在这些边界内,大量生物、化学和物理过程结合在一起,形成水、碳和其他营养元素的循环,并在整个系统中获取、转移和耗散能量(Loreau, 2010)。生态系统动力学的研究涉及生态系统的干扰或扰动(自然和人为)以及其在时间和空间上的影响(Turner et al., 2003; Chapin et al., 2011)。干扰可能包括短期事件,如野火、疾病、风暴和洪水,也可能包括长期过程,如降雨和温度变化,消除原生植被和栖息地破碎化(White, Pickett, 1985; Turner et al., 2003)。在某些情况下,自然干扰强度和频率的变化,如火灾间隔和火灾强度的增加或减少,足以引起生态系统结构和功能发生重要变化(Turner et al., 2003; Edwards et al., 2015)。

专栏 5.1

什么是生态系统过程、功能和服务?

尽管"生态系统过程、生态系统功能和生态系统服务"这些词被广泛用在生态学、环境科学和经济学中,但是它们之间的含义和相互关系往往界定不清。我在这里提供一些明确的定义。

生态系统过程(ecosystem process):当生态系统中的两个或多个部分,相互作用时发生的过程(Odum, 1983)。例如,包括日照(阳光照射在岩石或树木上)、侵蚀(快速流动的水将土壤从一个地方移动到另一个地方)、光合作用(植物将太阳光的能量转化为化学能)、固氮(豆科植物的根系内的细菌将大气中的氮转化为氨)、食草(动物吃植物)和捕食(动物吃动物)。生态系统中的重要循环,如碳循环和水循环,由许多不同的过程组成。

生态功能(ecosystem function):生态系统执行的功能,也即是生态系统发挥的功能。生态系统功能依赖于一个或多个生态系统过程并从中产生。例如,包括初级生产(依赖于降水、日照、光合作用以及影响土壤中生物可以利用的营养元素浓度的各种过程)、水文流动(依赖于降水、渗透、蒸散、渗滤等)和授粉(这依赖于将花粉从花的花药转移到柱头的各种过程,如风或鸟类、昆虫和哺乳动物不断地造访同一物种的花)。一些作者似乎交替使用"生态系统过程和生态系统功能"这两个术语(Costanza et al.，1997；Diaz，Cabido，1997；Crutsinger et al.，2006)。

生态系统服务(ecosystem service)：人类直接或间接地从生态系统功能中获得的利益(Costanza et al.，1997；Millennium Ecosystem Assessment，2003)。例如植物食品(植物体、果实和种子)、动物性食品(来源于次级生产，又依赖于初级生产等)、建筑用材(初级生产)和供应清洁水(来自流域中水的收集和过滤以及在湖泊和溪流中存储的水)。一些作者似乎将"生态系统功能和生态系统服务"交替使用，这意味着所有生态系统功能都以某种方式使人类受益(de Groot，1987；Luck et al.，2003；Tscharntke et al.，2005)。

像种群生态学和群落生态学一样(请参阅第3章和第4章)，生态系统生态学的概念和理论在很大程度上已被发展来描述和扩展我们对非城市生态系统的理解，例如森林、草地、湖泊、海洋或沙漠(Groffman et al.，2004)。然而，这些概念和理论中有许多已成功应用于城市系统(Pickett et al.，2001；Grimm，Redman，2004；Groffman et al.，2004；Grimm et al.，2008a、2008b；Pickett，Grove，2009)。在某种程度上，几乎所有生态系统都是开放系统，它们依赖于外部太阳能的输入(Odum，1983)。但是，城市生态系统的特征往往是通过人为途径从系统外部大量输入非太阳能、水和养分，然后再输出到周围的土地，接收水和大气也通过人为途径(Kennedy et al.，2007)。同时，城市生态系统中内源能量、水、碳和其他养分的利用和循环大大减少。在本章中，我考虑了城镇化的生物物理过程，是如何从局部到全球一系列空间尺度上影响生态系统内的重要循环。我还讨论了缓解这些影响的策略，例如增加城市中内源能量、水和养分的使用，并恢复各种生态过程以促进内部循环。

5.2 碳

5.2.1 碳循环的介绍

全球碳循环，是由许多不同空间尺度上的碳循环组成。碳通过多种生物地球化学过程中在大气、生物、岩石、沉积物、土壤、海洋、河流和湖泊之间交换。在光合作用过程中，植物和蓝藻将来自太阳光能转换为化学能，以每年（123±8）PgC 的速度从大气中提取二氧化碳（95%置信区间：102~135PgC/a，其中 1PgC = 10^{15}g 碳；Beer et al., 2010），并在此过程中释放氧气。产生的化学能以碳水化合物分子形式存储，由二氧化碳和水合成，然后用于生长或为细胞过程提供能量。在陆地生态系统中，这种有机体通过食物链，从植物向上移动到食草动物、食肉动物和捕食者，然后移动到以分解动植物体为食物的食腐动物和腐生物。据统计，1950~3050Pg 的有机碳存储在陆地生物圈中，该生物圈包括动物、植物和其他生物，活的和死的、地上和地下土壤中（Batjes, 1996；Prentice et al., 2001）。在湿地和多年冻土中，储存了约 2000Pg 的有机碳（Bridgham et al., 2006；Tarnocai et al., 2009）。土壤还以碳酸钙等形式储存一些无机碳。有机碳通过呼吸作用从土壤释放到大气中，而有机碳和无机碳均可从土壤中溶解并流入河流和海洋。

水生生态系统也在全球碳循环中起着重要作用，特别是通过大气和海洋之间的碳快速交换。大气中的二氧化碳溶解在海洋表层，然后转化为碳酸盐（溶解的无机碳），在光合作用中被海藻和其他植物转化为有机碳。这种有机碳在海洋食物链中移动，可以通过海洋生物的分解和呼吸作用进行循环利用，也可以沉积到海洋深处，在那里长时间储存后再循环到表层（Ciais et al., 2013）。有机碳还通过海洋生物的呼吸释放到大气中。海洋的中层和深层估计储存了 37000Pg 的溶解无机碳，而表层储存了约 900Pg（Ciais et al., 2013）。海洋中溶解的有机碳占 700Pg，而海洋生物包含了 3Pg 的有机碳（Hansell et al., 2009；Ciais et al., 2013）。目前，地球大气层储存的碳约为 832Pg。

在最近的几个世纪中，人类活动包括燃烧化石燃料（煤、石油和天然气）、砍伐森林和生产水泥增加了地球大气中碳的浓度（图 5.1a）。这些额外的大气碳，主要以 CO_2 和甲烷（CH_4）的形式存在，它们被称为温室气体，因为它们在对流层（大气的最底层）中吸收并释放热辐射，从而增加了地球的温度（Falkowski et al., 2000；IPCC, 2014）。虽然约 60%的人为 CO_2 排放已经被海洋和陆地生物圈吸收，但是其余部分仍留在大气中（CSIRO, BoM, 2014；图 5.1b）。

自从 1750 年工业革命以来，全球大气中 CO_2 的浓度增加了 40%以上，而甲

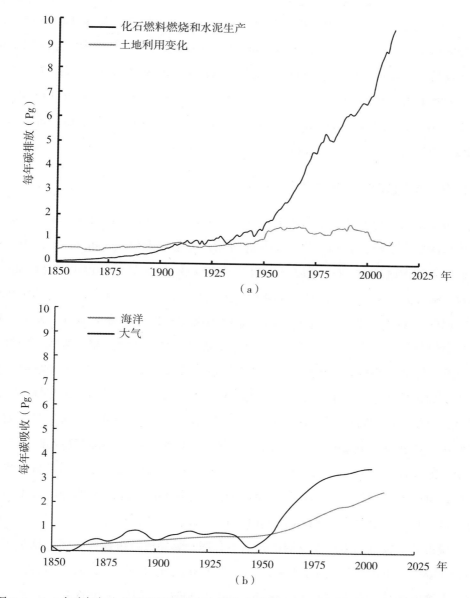

图 5.1 （a）全球每年由化石燃料燃烧和水泥生产（黑线）、土地利用变化（灰线）而导致的人为碳排放；（b）大气（黑线）和海洋（灰线）每年吸收的碳排放量(Le Quéré et al., 2015)

烷的浓度增加了150%(Ciais et al., 2013；Dlugokencky, Tans, 2015)。这极大地改变了地球的气候，包括空气和海洋的温度升高（全球变暖）、降雨模式的改变，以及干旱、洪水和飓风等极端天气事件的发生频率和严重性增加(CSIRO，BoM，2014；IPCC，2014)。人为碳排放也对海洋产生了重大影响。海水中碳浓度的增加会提高海水酸度，对钙化的浮游生物、珊瑚和软体动物等海洋生物造成了潜在的严重后果(Doney et al.，2009；Hoegh-Guldberg，Bruno，2010)。自1750年以来，公海中表层水的pH值下降了0.1，相当于氢离子浓度增加了26%(CSIRO，BoM，2014)。人为造成的CO_2排放量继续增加，估计2013年全球CO_2排放量为36Pg，相当于(9.9±0.5)PgC(Le Quéré et al.，2015)。

5.2.2 城镇化对碳循环的影响

城镇化以多种方式影响碳循环(Grimm et al.，2008a，2008b)。城市是能源密集型的生态系统，其CO_2排放量与能源输入非常接近(Kennedy et al.，2007)。据估计，城市区域的CO_2排放量占世界人为CO_2排放量的80%以上(Grubler，1994；Svirejeva-Hopkins et al.，2004)。鉴于目前全球约有54%的人口居住在城市中(UN，2004)，按人均计算，这一比例很高。在世界上许多地方，发电厂燃烧化石燃料为城市供电。这些电力供应给家庭、工业、办公大楼、医院、路灯和公共交通系统(如火车、电车)。人员和货物通过交通如公路、海运、空运进出城市及周边地区，这也依赖化石燃料的燃烧，而也通常燃烧石油、天然气、煤或木材，以供冬季取暖(Kennedy et al.，2009)。电力产生的温室气体排放量，取决于所用电量和单位电能所产生的温室气体排放量(称为温室气体强度或排放因子，以单位能源释放的CO_2eq为单位)。例如，南非开普敦的大部分电力来自燃煤发电厂，每一度(千瓦时)电将产生969g CO_2 eq；瑞士日内瓦的大部分电力来自水力发电，每一度电会产生54g CO_2eq(Kennedy et al.，2009)。

城市人口数量的增加以及城市规划不合理，会导致城市横向扩张，增加了通勤者在郊区和市内之间的出行距离。在不断扩张的城市中，广铺式的城市形态与较高人均能源使用是相关的(Kennedy et al.，2007)。例如，2005年，科罗拉多州丹佛市(人口密度为1558人/km^2)人均CO_2排放量为21.5Mg，而人口更密集的纽约市(人口密度为10350人/km^2)人均CO_2排放量为10.5Mg(Hoornweg et al.，2011)。对来自10个城市的人均温室气体排放量的回归分析显示，排放量随着人口密度的增加而下降(Kennedy et al.，2009；图5.2)。然而，丹佛的CO_2排放量较高的原因是冬季较冷,因此每年的供暖天数高于纽约(Kennedy et al.,2009)。城市的扩张也往往侵占郊区的农业用地,这增加了城市居民从生产地点到消费地点的食物运输距离(Pauchard et al.,2006;van Veenhuizen,2006;Kissinger,2012;Mok et al.,2014)。

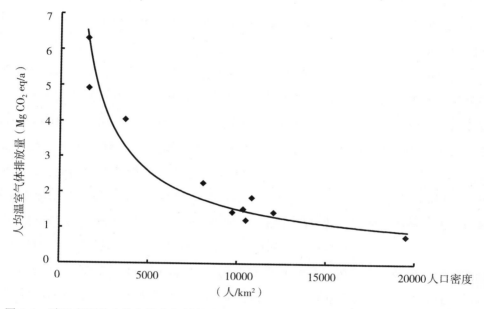

图 5.2 陆地交通的人均年温室气体排放量(以 $MgCO_2$ eq/a 计)与城市人口密度的对比(修改自 Kennedy et al.,2009,图 3)

现代城市的建设和扩张,驱动着全球用于建造混凝土建筑、道路、人行道、水坝、隧道、下水道雨水排水系统的水泥需求。水泥制造设施加热碳酸钙以产生石灰和 CO_2。这些 CO_2 被释放到大气中,而化石燃料的燃烧产生的能量用于制造水泥,这样会释放更多的 CO_2。结果是,水泥工业造成全球约 5%(3%~8%)的人为 CO_2 排放量,每生产 1Mg 水泥排放近 0.9Mg 的 CO_2(Mahasenan et al.,2003;Benhelal et al.,2012)。由于中国密集的城镇化进程以及混凝土在建筑中的广泛使用,中国在水泥熟料和混凝土生产中的 CO_2 排放量超过了其他所有国家(Fernandez,2007)。在 2005 年,世界混凝土总产量为 2.3Pg(相当于释放约为 2Pg 的 CO_2),其中 46% 由中国生产(Fernandez,2007)。钢材是另一种高耗能的建筑材料，每生产 1Mg 的钢材平均会排放 1.8Mg 的 CO_2（世界钢铁协会，2013）。钢材用于建造住房、商业建筑和城市基础设施，如铁路线、桥梁、电网、太阳能电池板和风力涡轮机；它也用于加固混凝土。2010 年，钢铁行业造成约占 6.7% 的全球 CO_2 总排放量（国际能源署，世界钢铁协会转引，2013）；2014 年，全球粗钢产量达到 1.64Pg，相当于释放 2.95Pg 的 CO_2（世界钢铁协会，2014）；在水泥和钢铁生产量方面，中国居世界领先地位，2014 年占全球产量的 50%（世界钢铁协会，2014）。

城市扩张将森林、草地和农业景观转变为城市景观，去除了原本可以从大气中吸收 CO_2 的植被，同时用沥青和混凝土覆盖在土壤上。这种转换会产生大量的碳排放，2010 年全球碳排放量估计为 158Tg（$1Tg = 1 \times 10^{12}g$）（Svirejeva-Hopkins et al.，2004）。非洲、中国和亚太其他地区的城市增长为这一数字作出了最大贡献。在 20 世纪 90 年代中期，美国由于城镇化而失去的生产性农业用地，使植物光合作用固定的碳每年减少 40Tg（Imhoff et al.，2004）。预计在未来几十年中，城市土地转换产生的碳排放量将进一步增加（Svirejeva-Hopkins et al.，2004）。但是，一旦发生城市土地转换，有机碳便会在有管理的区域的植被和土壤中积累，例如公园、花园、行道树、草坪和高尔夫场地（Pataki et al.，2006）。尽管需要进一步研究，但是迄今为止的结果表明，城市土地转换对城市碳平衡的净效应，部分取决于所替代生态系统的类型。例如，干旱、沙漠环境的城镇化可能会增加植被和土壤总储存的碳，而湿润的森林环境的城镇化则可能相反（Golubiewski，2003；Pouyat et al.，2003；Pataki et al.，2006）。

城市及其地区大气中的 CO_2 和 CH_4 浓度升高（Idso et al.，1998；Kaye et al.，2006；George et al.，2007；Grimm et al.，2008a）。主要由化石燃料的局部燃烧引起的城市区域内的 CO_2 累积，被称为 CO_2 穹顶效应（Idso et al.，2001）。根据气象条件，CO_2 穹顶可能会加剧当地的城市热岛效应（Balling et al.，2001；George et al.，2007）。但是，亚利桑那州凤凰城的一项研究发现，整个城市的 CO_2 穹顶效应仅占整个城市热岛效应的一小部分（Balling et al.，2001）。1950—2005 年，人为碳排放造成的全球变暖和城市热岛效应共同导致澳大利亚墨尔本的最低、平均和最高气温每十年分别增加 0.32℃、0.23℃和 0.14℃（Suppiah，Whetton，2007），并将每年的平均霜冻天数从 12.2 天（1856—1950 年）减少到

3.1天(1951—2000年)(图5.3;Parris,Hazell,2005)。城市温度升高增加了在夏季建筑物中使用空调的需求,这反过来增加了电力消耗和化石燃料的燃烧,形成了一个正反馈循环(Baker et al.,2002)。

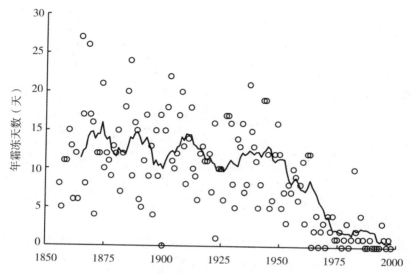

图5.3 澳大利亚墨尔本1850—2000年的年霜冻天数(圆圈)和年霜冻天数每10年平均变化曲线(改编自Parris,Hazell,2005,图3b)

5.2.3 缓解策略

减轻城镇化对全球碳循环的影响,最有效的方法是减少与城市建设、扩张和运行相关的大气碳排放。这可以通过减少为供应城市电力和建筑材料,以及为运输人员和货物提供动力而燃烧的化石燃料使用量来实现;尽可能减少用于城市发展而损失的生产性林地和农业用地的数量。尽管从概念上讲很简单,但是这些解决方案很难在实践中实现。《联合国气候变化框架公约》下的第一项国际条约《京都议定书》于1997年通过,并于2005年生效(UN,1998)。加入该协议的38个发达国家致力于减少温室气体排放:到2012年(第一个实施阶段;UN,1998),将温室气体排放量平均降低到1990年的5%,其中21个国家承诺到2020年将排放水平至少降低到1990年的18%(第二个实施阶段;UN,2012);2015年12月,在巴黎举行的联合国气候变化大会上有 195 个国家通过了首个全球气候协议。该协议旨在到 21 世纪下半叶，将人为产生的温室气体净排放量减少到零，并且如果至少有 55 个国家签署（这些国家温室气体排放量占全球温室气体排放量的 55%以上），该协议将具有法律约束力（UN，2015）。

城市可以通过减少电力和运输燃料的需求来减少温室气体排放，例如，通过提高工业效率、减少家庭人均能源使用量，以及减少人与货物的公路运输，而采用铁路和其他公共交通系统，从而减少温室气体的排放。城市人口密度的增加预计将减少人均能源使用量，特别是结合有效的公共交通系统和当地规划措施，可以减少在日常活动中对私人车辆的依赖（Kennedy et al.，2009）。城市还可以通过降低发电产生的温室气体排放强度来减少温室气体排放。燃煤发电在整个生命周期中的平均温室气体强度为 1001g CO_2eq/(kW·h)；相比之下，天然气、太阳能电池板、核反应堆、风力涡轮和水力发电的 CO_2eq/(kW·h) 分别为 469g、46g、16g、12g 和 4g（表 5.1）。

依靠煤炭发电的城市，如澳大利亚、中国、南非和美国等国家的许多城市，可以通过增加其他电力能源的比例（包括太阳能和风能等可再生能源），来大幅度减少 CO_2 排放量。尽管核能的温室气体排放强度较低，但它不是可再生能源。由于核电厂的事故会使工人、公众以及周围的陆地、水生生境暴露于危险水平的放射性环境中，因此，它也对人类健康和环境构成威胁；此类事故平均每八年发生一次（Maurin，2011）。水力发电需要修建大坝，这些大坝可能会对淡水、陆地环境以及当地居民的生活产生重大负面影响。如果大坝倒塌，也会对人类安全构成威胁，从而造成破坏性洪水。所有电力生产方式都会对人类、其他动物和环境产生影响（Brook，Bradshaw，2015）；未来的挑战是选择现有的方法和/或开发具有降低温室气体强度及对环境、人类社会其他方面的不利影响最小的新方法。

减少碳排放的地方行动包括采用绿色建筑标准，这既可以减少建筑物内含能源需求（如通过回收或使用低能耗材料，如泥砖、板岩和石材），也可以减少为供暖、制冷和照明的能源使用量（如无源太阳能设计，改进的隔热、双层的玻璃）(Reddy，Jagadish，2003；Yudelson，2007；Hammond，Jones，2008；Chua，Oh，2011)。许多绿色建筑标准或星级评定系统还鼓励当地充分使用太阳能或风能，以供应家庭用电和热水（Kajikawa et al.，2011）。对泰国的三种房屋风格进行生命周期评估发现,木屋对温室气体的影响为 18kg CO_2eq/m^2,而混凝土建筑为 209kg CO_2eq/m^2。提议中采用纤维水泥板作为墙体的房屋经计算对温室气体的间接影响为 140kg CO_2eq/m^2(O'Brien,Hes,2008)。马来西亚的绿色建筑倡议已促进建造了许多节能办公楼,如能源部、绿色技术和水利部在布城的低能耗办公楼,以及在马来西亚绿色技术公司在班达尔巴鲁班吉的绿色能源办公楼(Chua,Oh,2011)。这两栋建筑的建筑能耗指数(BEI)分别为 100kW·h/(m^2·a)和 35kW·h/(m^2·a),而马来西亚传统办公楼的平均建筑能耗指数为 250kW·h/(m^2·a)(Chua,Oh,2011)。

表 5.1 通过对各种发电技术的温室气体排放的生命周期进行回归分析得出各种发电技术的温室气体排放强度

值	各种发电技术的温室气体排放强度[g CO_2eq/(kW·h)]										
	生物能	太阳能 PV	集中太阳能 CSP	地热	水力发电	海洋能	风能	核能	天然气	石油	煤
最低	-633	5	7	6	0	2	2	1	290	510	675
25%	360	29	14	20	3	6	8	8	422	722	877
50%	18	46	22	45	4	8	12	16	469	840	1001
75%	37	80	32	57	7	9	20	45	548	907	1130
最高	75	217	89	79	43	23	81	220	930	1170	1689
CCS 最小	-1368								65		98
CCS 最大	-594								245		396

注:CCS—碳捕获和碳储存,PV—光伏,CSP—集中太阳能(数据来源:Moomaw et al.,2011)

城市植被和土壤在碳封存方面发挥着重要作用,将这些碳从大气中吸收并储存在地上和地下生物量中(Jo,McPherson,1995;Nowak,Crane,2002;Pataki et al.,2006;Stoffberg et al.,2010)。树木、花园、屋顶花园和垂直绿化还可以通过遮阴和增加蒸散量来缓解全球变暖的局部影响,以及温带、热带地区中城市热岛效应(Tyrvainen et al.,2005;Roth,2007;Norton et al.,2013)。参考美国的七个城市，通过隔离和避免相结合一棵街道或公园中的树每年可以减少 31~252kg 的 CO_2 排放量。据估计，在整个城市范围内，亚利桑那州的格伦代尔市和明尼苏达州的明尼阿珀利市的街道和公园树木分别将城市的 CO_2 排放量减少了 0.7Gg 和 50Gg($1Gg=10^9$ g)(McPherson et al., 2005; Pataki et al., 2006)。按面积计算，格伦代尔市的 CO_2 排放量每年减少了 45.8kg/hm²；明尼阿珀利市的 CO_2 排放量每年减少了 4118kg/hm²。维持或增加具有透水土壤的绿色开敞空间，在街道、公园和私人花园中种植阔叶树可以提供夏季遮阴，在现有建筑物上创建绿色屋顶花园和垂直绿化，并用雨水或雨水径流灌溉这些不同类型的植被，让它们积极生长，从而在干旱天气中可以吸收 CO_2(详见 5.3.3 小节)，这都将有助于实现碳封存和局部降温的双重目标(Oberndorfer et al., 2007; Norton et al., 2013, 2015)。例如，在加利福尼亚伯克利，由于夏季街道树木为建筑遮阴，据估计每棵树可以节约 95kW·h 的电量(整个城市的用电量约为 3.5×10^8 kW·h(McPherson et al., 2005)。

城市植被的其他好处包括可以去除大气中的污染物(如臭氧、一氧化碳、二氧化硫、二氧化氮和固体颗粒物)，可以吸收污染土壤中的污染物，以及为当地生物多样性提供栖息地(Dwyer et al., 1992; Dimoudi, Nikolopoulou, 2003; McPherson et al., 2005; Nowak et al., 2006; Oberndorfer et al., 2007)。城市树木的一个潜在缺点是某些树种会产生植物源挥发性有机化合物(BVOCs)，如果其他前体污染物的浓度足够高，则可能导致地表臭氧的二次形成(McPherson et al., 2005; Calfapietra et al., 2013)。这个问题可以通过种植低或无 BVOCs 的树种来解决(Calfapietra et al., 2013)。根据种植的树种，城市植被还可以促进城市当地的粮食生产。都市农业(Urban Agriculture)减少了食物从生产地点到城市居民的距离(食物里程)，从而减少了在冷藏和运输中使用的化石燃料(Smit, Nasr, 1992; McClintock, 2010)。都市农业在澳大利亚、英国和美国等发达国家越来越受欢迎(Lovell, 2010; McClintock, 2010; Mok et al., 2014; 图 5.4)，而在非洲、亚洲和东欧许多地区，都市农业也已经成为城市居民重要的新鲜食物来源(Mougeot, 2005)。古巴在都市农业方面处于世界领先地位，自 1994 年以来，都市农业产量增长了 1000 倍(Koont, 2011)。苏联解体后，古巴无法获得大规模的工业化农业所需要的化石燃料和化肥。因此，它将农业项目的重点聚焦于城市和周边废弃公共土地上进行可持续有机农业生产，并取得了巨大的成功。

图 5.4　英国约克花园的私人菜地（摄影：Kirsten M. Parris）

5.3　水

5.3.1　水循环介绍

　　水在大气、生物圈、地面、海洋、冰原和冰川，以及液态淡水（小溪、河流和湖泊等）之间不断循环运动。像碳循环一样，水循环包括许多在一定时空范围尺度上运行的过程。太阳的热量导致水的蒸发并以水蒸气的形式进入大气。大气中的水蒸气随气流移动，上升、冷却和下降成为降水（雨、雨夹雪或雪）。大部分降落在土地上的雨水都渗入土壤中，并以土壤水分的形式储存，通过地下流渗流到溪流、池塘或湖泊中，或进入地下蓄水层；一小部分降水以地表径流流入河流和其他水体。据估计，全球年平均降水量约为 $0.5 Pm^3$（$1 Pm^3 = 10^{15} m^3$）

(Pidwirny，2006)。其中，大约80%的降水落在海洋上，其余20%落在陆地上。在陆地环境中，植物从土壤吸收水分，并通过蒸散过程将其转移到大气中。在全球范围内，据估计植被覆盖的土地上的蒸散量为每年62.8Tm^3，超过陆地降水量的60%。小溪和河流流入大海，降水又返回海洋，而湖泊和地下蓄水层则可能储水时间更长。在融化或升华(直接从固态水转化为水蒸气)之前，水还可以以冰和雪的形态长期保存(长达数十万年)。

5.3.2 城镇化对水循环的影响

城镇化通过土地覆盖变化(如去除现存植被；建造建筑物、道路和其他城市基础设施；用坚硬的表面和带衬里的排水管替代可渗透的表面和径流路线；缩减开敞空间的面积；对水生生境的改造或破坏)及产生污染和废物而影响水循环。不透水表面阻止雨水渗入土壤(Stone，2004)。当与高效的排水系统相结合时，这些水会迅速输送到小溪、湖泊和河流中(Leopold，1968；Walsh et al.，2005)。临时池塘被破坏和自然排水管线的改造(如用混凝土或金属管道代替短暂的一级河流，或把高阶河流改造成内衬混凝土的通道)会增加降雨后的水流速度，使问题更加复杂化(Lindh，1972；Walsh et al.，2005)。另外，去除植被和失去开敞、可渗透的空间，会降低城市的蒸散速率，这不是由树木和其他植物从土壤中吸收水分，然后释放到大气中，而是变成较少的可渗透表面以及较少的植物去转移水分(Kondoh，Nishiyama，2000；Walsh et al.，2012)。

城市环境中的这些变化增加了当地侵蚀性水流的频率和幅度，增加了城市溪流的峰值流量，缩短了降雨事件和洪峰流量到达之间的滞后时间，即使在降雨量相对较小的情况下，也会造成暴雨水文曲线(城市河流的"暴洪")快速上升和下降(Paul，Meyer，2001；Stone，2004；Walsh et al.，2005；见第2章)。将整个城市的生态系统作为一个整体，饮用水、生活用水和工业用水的使用，然后其中大部分通过污水(废水)系统排出，以及雨水经雨水系统从城市环境中快速排出，这往往主导城市的水流(图5.5)。这破坏了城市水循环的周期性，取而代之的是城市充当大量水的进入和排出的双重作用：最显著的是饮用水流入城市，然后通过污水处理系统再次流出；雨水落在城市上，然后通过雨水系统流出。

在大多数现代城市的正规居住区中，污水是由中心处理厂进行收集、处理，

5.3 水

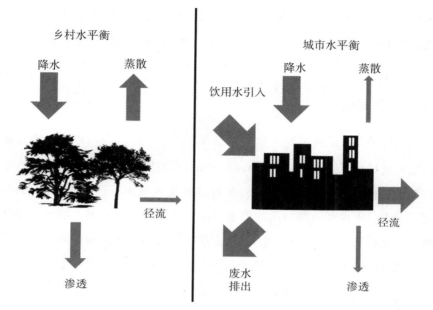

图 5.5 城乡水平衡示意图。在城市中，单位面积的蒸散率和渗透率低于乡村生态系统，而径流量高于乡村生态系统。城市生态系统也往往以大量饮用水的引入和大量废水的排出为主要特征

从而减少了向河流和沿海水域排放营养物质、化学物质、微生物病原体和其他污染物的负荷。然而，城市雨水径流流入溪流和其他未经处理的受纳水域，并携带大量污染物，包括氮、磷、石油产品、重金属和宠物粪便污染物（Lee et al.，2002；Hatt et al.，2004；Walsh et al.，2005，2012；Li et al.，2007；Imberger et al.，2011）。应当指出的是，在欧洲和北美洲的一些较旧的城市中，污水和雨水系统之间没有很清楚地分开，从而加剧了降雨后受纳水体的污染事件（Grimm et al.，2008a）。在有大量非正式居住区或贫民窟的城市中，城市水污染问题更为严重。非正规住区通常在没有规划控制的情况下建造的，有时是在受洪水影响的地区，缺乏管理、应对暴雨的基本的卫生设施和基础设施。降雨后，这些条件导致雨水径流、人类排泄物和家畜的粪便混合在一起，传播各种病原体，有引发疾病的风险，并且污染了受纳水体（Jagals，1996；Parkinson，2003；Owusu-Asante，Stephenson，2006；Owusu-Asante，Ndiritu，2009）。这些受污染的水源，会经常

作为居民的其他用水，包括饮用、渔业和农作物灌溉等用途(Jagals，1996；Parkinson et al.，2007)。开敞的排水沟和水塘受到污染，也对非正规居住区的居民构成了严重的健康风险，并为传播疾病的蚊子和寄生虫提供了温床(Parkinson，2003)。

5.3.3 缓解策略

城市水管理面临的挑战是来自难以管理饮用水、废水和雨水，以实现多个目标，甚至有时是相互竞争的目标(Grimm et al.，2008a)。有效地管理流入指定处理厂的封闭式管道或下水道的废水，并有效地疏排城市不透水表面雨水的径流，从而达到卫生、疾病预防和防洪的多重目标，以及将城市街道上的水污染物有效地转移至溪流和其他受纳水体中。但是，这些做法也会降低城市河流的水质，将污染物转移到河流下游，减少了保留和使用雨水的机会。这些雨水是可以改善城市环境的，例如支持植被系统如公园、行道树、屋顶花园等，从而反过来提供一系列的生态系统服务，如遮阴、微气候降温、吸收空气污染物及为本地生物多样性提供栖息地(Paul，Meyer，2001；Wong，2007；Fletcher et al.，2007；Ashley et al.，2013；Fletcher et al.，2014)。

近几十年，城市用水管理越来越复杂，既定目标也越来越多(图 5.6)。在发达国家，人们越来越意识到城镇集水区保留雨水的生态效益和社会效益，催生了一种新的城市水管理方法，该方法被称为水敏城市设计(澳大利亚)、低影响设计、绿色基础设施(美国)、可持续城市排水系统(英国)和技术替代方案(法国)等(Fletcher et al.，2014)。水敏城市设计及相关实践旨在保护和改善城市地区的自然水系统；将雨水处理纳入景观；通过当地的滞留雨水措施减少雨水径流和洪峰流量，并降低不透水表面与受纳水体之间的连通性；并改善城市区域的排水水质(Fletcher et al.，2014)。这些设计采用了一系列的工程和计划措施，以减少流入城市溪流的雨水量，降低雨后的径流速度以及雨水携带污染物的负荷(Wong，2007)。例如：安装水箱，修建雨水花园、可渗透路面、屋顶花园、蓄水池、人工湿地和土洼地(Walsh et al.，2005；Mentens et al.，2006；Hilten et al.，2008；Beecham et al.，2012)。

尽管水敏城市设计原则可能适用于所有类型的城市，但在许多非正规居住

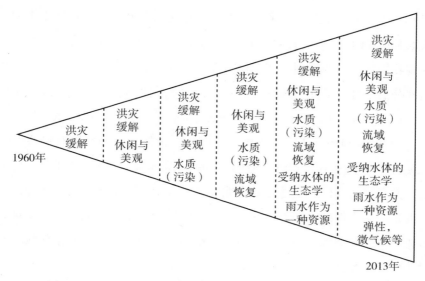

图 5.6 城市水管理的目标和时间（Fletcher et al.，2014，图 2）

区，目前的水资源管理的重点仍然是防止洪水泛滥，以及有效地将废水与雨水分离（Parkinson et al.，2007；Okoko，2008；Owusu-Asante，Ndiritu，2009）。在解决这些问题之前，管理目标（如恢复自然流域、受纳水域的生态系统以及休闲、美观）不太可能具有太大的重要性（图 5.6）。上面概述的一些城市水管理技术适用于贫民窟和其他非正规居住区，特别是使用渗透路面和收集雨水作为居民的饮用水（Handia et al.，2004；Cowden et al.，2006；Owusu-Asante，Ndiritu，2009）。但是，这些雨水被人类废弃物污染时，通过滞留池、人工湿地和沼泽等结构将雨水保留在城市集水区的可取之处就不存在了——任何确保这种水从城市环境中缓慢排放的措施，都会给人的健康构成风险（Parkinson，2003）。如果可以在一个单独的封闭系统中有效地处理人类废弃物，那么采取一系列的工程措施来减少非正规居住区的雨水径流量和养分负荷可能是既实用又非常有利的。

5.4 氮循环

已知有 17 种营养物质对植物的功能和生长必不可少：碳、氢、氧、氮、磷、钾、钙、镁、硫、硼、氯、铜、铁、锰、钼、镍和锌（Barker，Pilbeam，2015）。

植物以 CO_2 的形式从大气中吸收碳,并从水中吸收氢和氧。它们从土壤的矿物成分中获取其他 14 种必需的营养物质。植物生长,即生态系统的初级生产力,常常受到营养物质供应的限制,其中包括大气中碳的浓度以及土壤中氮、磷和其他矿物质养分的浓度。因此,假设其他营养物质不受限制,大气中 CO_2 含量的增加可能会提高光合作用和植物的生长速度(Wang et al., 2012; Reich, Hobbie, 2013)。氮和磷的利用率还影响土壤、腐殖质、枯枝落叶和水中有机质的分解速率(Flanagan, Van Cleve, 1983; Aerts, 1997; Manzoni et al., 2010)。每种必需营养物质都有其自身的生物地球化学循环,本章中没有足够的篇幅来详细地讨论它们。但是,每个循环周期都可能受到城市建设和扩张的影响。随着全球人口日益城镇化,粮食和能源资源将越来越向城市集中。因此,预计城市区域将在全球营养物质分配中发挥着重要作用(Bernhardt et al., 2008)。

氮是所有生物最重要的营养物质之一,通常会限制植物的生长和初级生产力(Vitousek et al., 1997; Dentener et al., 2006; Gruber, Galloway, 2008; Zehr, Kudela, 2011)。氮是蛋白质和核酸 DNA、RNA 的基本成分,存在于每个活细胞中。植物和蓝藻也需要氮才能产生叶绿素,叶绿素在光合作用中起关键作用。氮(N_2)是地球大气中的主要气体,是全球循环中最大的氮库。但是,这种不活泼的氮无法被植物吸收,在没有人为干预的情况下,首先必须由固氮菌将氮转换为铵根(NH_4^+)或通过闪电转化为氮氧化物(NO_x)。这些过程称为固氮。固氮菌要么以自由生物的形式存在(例如海洋或海洋沉积物中的某些厌氧细菌),要么与植物共生(例如豆科植物根瘤中的根瘤菌)。它们通过将气态 N_2 与氢结合产生氨来固定氮。雷暴期间,N_2 被闪电氧化,会生成活性氮氧化物,它们与羟基自由基(OH^-)结合形成硝酸(HNO_3),然后通过沉淀转移到地球表面的陆地、淡水或海洋环境中(Bond et al., 2002)。

在陆地生境中,植物通过根系吸收、同化铵或者硝酸根离子,将它们合成为植物功能和生长所需的有机化合物(如叶绿素、氨基酸和核酸)。反过来,动物通过消耗植物和其他动物来满足其大部分氮需求。当有机体死亡或动物排泄废物时,细菌或真菌会通过氨化过程将有机氮转化为氨。还有一些细菌将氨转化为亚硝酸盐,然后转化为硝酸盐(硝化作用)。另一类细菌在厌氧条件下(如在涝渍土壤中发现的),将硝酸盐转化为 N_2(反硝化作用),从而完成氮循环(Gruber,

Galloway,2008)。在海洋中的氮循环也发生类似的过程,其中蓝绿藻(蓝细菌)负责大部分的氮固定。在近海和开阔海域中,浮游植物对氮的吸收以铵或硝酸盐的形式,生物有效氮的浓度是初级和次级生产的主要驱动力(Gruber,Galloway,2008;Zehr,Kudela,2011)。

5.4.1 城镇化对氮循环的影响

人类活动,尤其是化石燃料的燃烧,通过哈伯法对氮进行工业固定(制造以氨为基础的肥料),以及种植以游离固氮及与共生细菌固氮作用的豆类和水稻等农作物,大大提升了氮的利用率(Galloway et al.,2004)。21世纪,全球人为固氮率(活性氮210Tg/a)比20世纪高出10倍,现在已经等于自然资源中的氮固定量(Fowler et al.,2013)。鉴于世界上城市人口、能源和营养物质的高度集中,大部分这种活性氮要么在城市环境中产生(燃烧化石燃料用于发电、供暖和车辆运输),要么以食物、花园肥料和工业化学品的形式人为地带入城市中(Bernhardt et al.,2008;图5.7)。

在基础设施充足、环境法规完善的城市中,大部分作为人类食品和工业化合物的输入氮,都在化粪池和污水处理系统中进行处理,或者作为固体废物在指定的垃圾填埋场进行处置(Bernhardt et al.,2008)。但是,对以宠物食品和花园肥料的形式进入城市的氮控制得不是很好。结果是宠物粪便和来自施肥草坪、花园的径流,对城市生态系统的氮平衡起到非常重要作用(Baker et al.,2001;Groffman et al.,2004)。例如,在亚利桑那州马里科帕县,估算分别有58%和14%的氮输入来自花园施肥和宠物废物,每年贡献的总氮为10.4Gg和2.4Gg(Baker et al.,2001)。车辆尾气以 NO_x 和 NH_3 的形式排放氮,而燃烧化石燃料用于电力生产、家庭取暖和工业生产时会产生大量 NO_x(van Aardenne et al.,2001;Bernhardt et al.,2008;Gu et al.,2012,2013)。大气中局部的、干湿的氮沉积是城市环境中有效氮的重要且不断增加的组成部分(Kennedy et al.,2007)。在中国,自1980年以来,化石燃料燃烧产生的 NO_x 排放量增加了6倍多,2008年估计为6Tg N(Gu et al.,2012)。据计算,全中国的平均氮沉积速率为12.9kg N/($hm^2 \cdot a$),各省之间的差异很大,具体取决于城镇化水平和人口密度[范围:新疆为3.8kg N/($hm^2 \cdot a$),至台湾的38.2kg N/($hm^2 \cdot a$)](Lu,

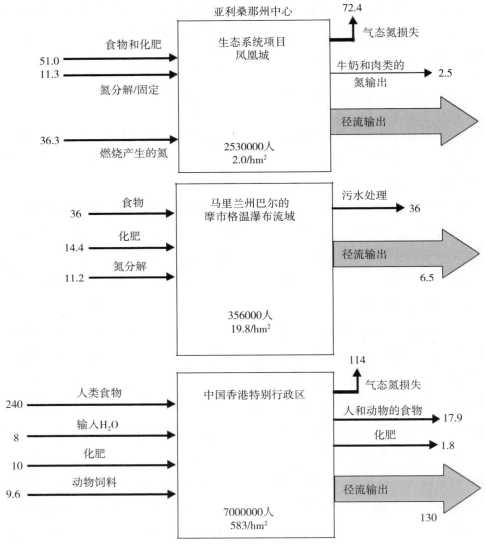

图 5.7 三个城市的氮质量平衡，显示了氮的输入和输出：美国亚利桑那州凤凰城，美国巴尔的摩的格温瀑布流域，中国香港特别行政区（Bernhardt et al., 2008, 图 2）

Tian，2007）。

城市中氮的利用率的提高对城市生态系统有许多影响。与周围乡村地区相比，在陆地城市环境中增加生物可利用氮的供应可能导致植物生长速度加快、初级生产力提高和生物量积累增加(Vitousek et al., 1997)。在一些地区，当地植物适应低氮土壤(如澳大利亚、南非、地中海、加利福尼亚州和其他沙漠生物群落)，额外的土壤氮素可能更有利于外来入侵植物而不是本地植物，导致维管植物群落组成变化和物种丰富度的丧失(Bobbink et al., 1998; Dukes, Mooney, 1999; Brooks, 2003; Daehler, 2003)。在芬兰北部奥卢附近的外生性真菌和地衣中也观察到类似的模式，随着城市污染的增加(包括土壤中氮含量的增加)，物种丰富度呈现下降趋势(Tarvainen et al., 2003)。

城市环境中氮的利用率增加，也会对淡水和海洋系统产生影响，有时会非常严重。高效的雨水排水网络，在降雨事件后，将氮从陆地城市环境迅速输送到河流、湖泊、河口和海洋等水域，污水排放(处理过的和未处理的)和受污染的城市大气中氮的沉积都导致水生生境中的氮含量升高(Rabalais, 2002; Bernhardt et al., 2008)。其后果包括初级生产力增加、富营养化(有机物积累)、浑浊度增加及水生植物的日照接受量相应减少，有害藻华、氧气耗竭、植物多样性丧失和鱼类种群崩溃(Rabalais, 2002; Camargo et al., 2005)。值得注意的是，某些溪流的生产力更多地受到磷的利用率而不是氮的限制，而城市流域流入的磷，对河流的诸如富营养化之类的过程的影响可能大于氮(Taylor et al., 2004)。硝酸盐对多种淡水动物来说是有毒的，包括无脊椎动物、鱼类和两栖动物；毒性随着硝酸盐浓度和暴露时间的增加而增加，但随着动物体型大小和盐度的增加而降低(Camargo et al., 2005)。推荐的硝酸盐安全浓度水平是：在淡水中 NO_3-N(硝态氮)为 2mg/L，在海洋系统中 NO_3-N 为 20mg/L(Camargo et al., 2005)。铵和硝酸盐的沉积也会增加淡水系统的酸度，进而降低鱼类的存活率、大小和多度(Rabalais, 2002)。最后，河流、河口和沿海水域的氮沉积使得这些水体向大气中释放 N_2O 的量增加(Seitzinger, Kroeze, 1998)。N_2O(笑气)是一种强力的温室气体，在 20 年内，其全球变暖潜力比 CO_2 大 200 倍以上(Ciais et al., 2013)。

5.4.2 缓解策略

有许多可行的策略可以减少城市和邻近环境中活性氮的利用率，其中一些策略比其他方法更难实施。例如，碳排放减排策略，包括通过减少为城市提供电力、运输、供暖和建筑材料而燃烧的化石燃料(特别是煤和石油)，从而减少城

市和邻近环境中 NO_x 和 NH_3 的排放(参阅 5.2.3 小节)。特别是在发展中国家,对燃煤发电厂的 NO_x 排放的严格环境监管,也可能对减轻全球 NO_x 污染产生重要影响。在 2010 年,中国 NO_x 排放量相对于 GDP 的比重为 3.4Mg/百万美元,而美国为 0.9Mg/百万美元,日本为 0.3Mg/百万美元(Hill,2013)。但是,基于此测算,土耳其和澳大利亚等其他国家的 NO_x 排放量也相对较高(分别为 2.3Mg/百万美元和 1.8Mg/百万美元)。通过对机动车的 NO_x 和 NH_3 排放实施更严格的标准,并调节大气中 NH_3 的浓度,可以进一步减少城市区域的氮沉积(Moomaw,2002;Bernhardt et al.,2008)。

扩大城市的卫生基础设施并采用先进的废水处理方法,将大大减少城市区域向受纳水体输出氮(Bernhardt et al.,2008)。例如,尽管在过去的 10 年中,罗马尼亚、塞普洛斯、斯洛文尼亚、冰岛和西班牙等国家取得了实质性的进展,但是欧盟 1/3 人口的居住地在 2005 年根本没有污水处理设施(Katsiri,2009)。采取措施减少雨水排放的速度和数量,并将水保留在城市景观中,也将减少向河流和其他受纳水体的氮输出量(详见 5.3.3 小节)。这些措施包括将雨水分流到家庭蓄水池,并将雨水保留在有植被的湿地、沼泽和池塘中进行反硝化(Hsieh et al.,2007;Bern-hardt et al.,2008)。可以通过准则或法规限制在运动场、草坪、公园和花园中使用无机氮肥,减少向城市环境中输入氮,同时制定更严格的指导方针和/或适当执行现有的措施,负责任地处理宠物废物,也将减少向城市和乡镇输入活性氮(Driscoll et al.,2003)。

5.5 总结

从地方生态系统的过程和功能,到水、碳和其他营养物质的全球循环,城市的建设、扩张和运营对一系列空间尺度的生态系统都具有重大的影响。随着时间的流逝,世界人口从乡村到城市的持续迁移将使全球能源、水、营养物质和其他资源的集中度不断增加。但是,可以通过改变生产方式和方法以减少城镇化对生态系统的不利影响,例如:改变大规模地燃烧化石燃料以用于电力生产、家庭供暖和建筑材料生产的做法;建造节能住房和办公楼;减少依靠私人车辆运输人员和货物;减少城市周边地区生产性农田的流失;减少或禁止从城市中清除植被;

改变导致雨水及其污染物快速输出至本地河流和其他受纳水体的做法；减少或禁止无机氮肥在城市公园和花园中的广泛使用。通过增加风能和太阳能等可再生资源生产的电力比例；严格管制化石燃料燃烧产生的污染物；改善公共交通系统；采用绿色建筑，城市绿化战略，水敏城市设计以及限制城市无序蔓延的措施；未来的城市将能够为更多的人提供支持，同时减少对生态系统过程、功能和服务的负面影响。在下一章中，我将探讨城市人口的生态问题，以及城镇化对人类健康和福祉的诸多影响。

思考题

1. 什么是生态系统？
2. 定义"生态系统过程""生态系统功能"和"生态系统服务"这些术语。它们之间有何联系？
3. 解释城镇化影响全球碳循环的主要方式，以及缓解这些影响的实用策略。
4. 你是否认为发达国家应负责减少其城市对全球碳和水循环的影响？
5. 概述水敏城市设计的原则。在正规和非正规居住区中，如何应用这些原则？
6. 讨论应该如何限制城市向外扩张，以保护生产性耕地，并减少人均能源使用量。
7. 描述过量氮对城市环境（陆地和水生环境）的生态影响。在你知道的城镇中，是如何减少这些影响的？

本章参考文献

Aerts R. (1997) Climate, leaf litter chemistry and leaf litter decomposition in terrestrial ecosystems: A triangular relationship. *Oikos*, 79, 439-449.

Ashley R, Lundy L, Ward S, et al. (2013) Water-sensitive urban design: opportunities for the UK. *Proceedings of the ICE-Municipal Engineer*, 166, 65-76.

Baker L A, Brazel A J, Selover N, et al. (2002) Urbanization and warming of Phoenix (Arizona, USA): impacts, feedbacks and mitigation. *Urban Ecosystems*, 6, 183-203.

Baker L A, Hope D, Xu Y, et al. (2001) Nitrogen balance for the Central Arizona-Phoenix (CAP) ecosystem. *Ecosystems*, 4, 582-602.

Balling R C, Cerveny R S, Idso C D. (2001) Does the urban CO_2 dome of Phoenix, Arizona contribute to its heat island? *Geophysical Research Letters*, 28, 4599-4601.

Barker A V, Pilbeam D J. 2015. *Handbook of Plant Nutrition*. 2nd ed. CRC Press, Boca Raton.

Batjes N H. (1996) Total carbon and nitrogen in the soils of the world. *European Journal of Soil Science*, 47, 151-163.

Beecham S, Kandasamy J, Pezzaniti D. (2012) Stormwater treatment using permeable pavements. *Water Management*, 165, 161-170.

Beer C, Reichstein M, Tomelleri E, et al. (2010) Terrestrial gross carbon dioxide uptake: global distribution and covariation with climate. *Science*, 329, 834.

Benhelal E, Zahendi G, Hashim H. (2012) A novel design for green and economical cement manufacturing. *Journal of Cleaner Production*, 22, 60-66.

Bernhardt E S, Band L E, Walsh C J, et al. (2008) Understanding, managing, minimizing urban impacts on surface water nitrogen loading. *Annals of the New York Academy of Sciences*, 1134, 61-96.

Bobbink R, Hornung M, Roelofs J M. (1998) The effects of airborne pollutants on species diversity in natural and semi-natural European vegetation. *Journal of Ecology*, 86, 717-738.

Bond D W, Steiger S, Zhang R, et al. (2002) The importance of NO_x production by lightning in the tropics. *Atmospheric Environment*, 36, 1509-1519.

Bridgham S D, Megonigal J P, Keller J K, et al. (2006) The carbon balance of North American wetlands. *Wetlands*, 26, 889-916.

Brook B W, Bradshaw C J A. (2015) Key role for nuclear energy in global biodiversity conservation. *Conservation Biology*, 29, 702-712.

Brooks M L. (2003) Effects of increased soil nitrogen on the dominance of alien annual plants in the Mojave Desert. *Journal of Applied Ecology*, 40, 344-353.

Calfapietra C, Fares S, Manes F, et al. (2013) Role of Biogenic Volatile Organic

Compounds (BVOC) emitted by urban trees on ozone concentration in cities: A review. *Environmental Pollution*, 183, 71-80.

Camargo J A, Alonso A, Salamanca A. (2005) Nitrate toxicity to aquatic animals: a review with new data for freshwater invertebrates. *Chemosphere*, 58, 1255-1267.

Chapin F S, Matson P A, Mooney H A. (2011) *Principles of Terrestrial Ecosystem Ecology*. Springer, New York.

Chua S C, Oh T H. (2011) Green progress and prospect in Malaysia. *Renewable and Sustainable Energy Reviews*, 15, 2850-2861.

Ciais P, Sabine C, Bala G, et al. (2013) Carbon and other biogeochemical cycles. In *Climate Change* 2013: *The Physical Science Basis. Contribution of Working Group I to the Fifth Assessment Report of the Intergovernmental Panel on Climate Change*. Stocker T F, Qin D, Plattner G K, et al. eds. Cambridge University Press, Cambridge, 465-570.

Costanza R R, d'Arge R, de Groot R S, et al. (1997) The value of the world's ecosystem services and natural capital. *Nature*, 387, 253-260.

Commonwealth Scientific and Industrial Research Organisation (CSIRO), Australian Bureau of Meteorology (BoM). (2014) *State of the Climate* 2014. Australian Government Bureau of Meteorology, Canberra. http://www.bom.gov.au/state-of-the-climate (accessed 23/06/2014).

Cowden J R, Mihelcic J R, Watkins D W. (2006) Domestic rainwater harvesting assessment to improve water supply and health in Africa's urban slums. In *Proceedings of the World Environmental and Water Resource Congress* 2006: *Examining the Confluence of Environmental and Water Concerns*. Graham R ed. American Society of Civil Engineers, Reston, 1-10.

Crutsinger G M, Collins M D, Fordyce J A, et al. (2006) Plant genotypic diversity predicts community structure and governs an ecosystem process. *Science*, 313, 966-968.

Daehler C C. (2003) Performance comparisons of co-occurring native and alien invasive plants: implications for conservation and restoration. *Annual Review of Ecology*,

Evolution, *Systematics*, 34, 183-221.

de Groot R S. (1987) Environmental functions as a unifying concept for ecology and economics. *Environmentalist*, 7, 105-109.

Dentener F, Drevet J, Lamarque J F, et al. (2006) Nitrogen and sulfur deposition on regional and global scales: A multimodel evaluation. *Global Biogeochemical Cycles* 20, GB4003, doi: 10.1029/2005GB002672.

Diaz S, Cabido M. (1997) Plant functional types and ecosystem function in relation to global change. *Journal of Vegetation Science*, 8, 463-474.

Dimoudi A, Nikolopoulou M. (2003) Vegetation in the urban environment: microclimatic analysis and benefits. *Energy and Buildings*, 35, 69-76.

Dlugokencky E, Tans P P. (2015) Recent Global CO_2, NOAA, ESRS. www.esrl.noaa.gov/gmd/ccgg/trends/global.html (accessed 02/07/2015).

Doney S C, Fabry V J, Feely R A, et al. (2009) Ocean acidification: The other CO_2 problem. *Annual Review of Marine Science*, 1, 169-192.

Driscoll C T, Whitall D, Aber J, et al. (2003) Nitrogen pollution in the Northeastern United States: sources, effects, management options. *BioScience*, 53, 357-374.

Dukes J S, Mooney H A. (1999) Does global change increase the success of biological invaders? *Trends in Ecology and Evolution*, 14, 135-139.

Dwyer J F, McPherson E G, Schroeder H W, et al. (1992) Assessing the benefits and costs of the urban forest. *Journal of Arboriculture*, 18, 227-234.

Edwards A, Russell-Smith J, Meyer M. (2015) Contemporary fire regime risks to key ecological assets and processes in north Australian savannas. *International Journal of Wildland Fire*. 10.1071/WF14197.

EEA. (2012) *Greenhouse Gas Emission Trends and Projections in Europe* 2012—*Tracking Progress Towards Kyoto and* 2020 *Targets*. A Report by the European Environment Agency (EEA). Publications Office of the European Union, Luxembourg. http://www.eea.europa.eu/publications/ghg-trends-and-projections-2012 (accessed 11/12/2012).

Falkowski P, Scholes R J, Boyle E, et al. (2000) The global carbon cycle: A test of our knowledge of Earth as a system. *Science*, 290, 291-296.

Fernandez J E. (2007) Resource consumption of new urban construction in China. *Journal of Industrial Ecology*, 11, 99-115.

Flanagan P W, Van Cleve K. (1983) Nutrient cycling in relation to decomposition and organic-matter quality in taiga ecosystems. *Canadian Journal of Forest Research*, 13, 795-817.

Fletcher T D, Mitchell V G, Deletic A, et al. (2007) Is stormwater harvesting beneficial to urban waterway environmental flows? *Water Science & Technology*, 55, 265-272.

Fletcher T D, Shuster W, Hunt W F, et al. (2014) SUDS, LID, BMPs, WSUD, and more—The evolution and application of terminology surrounding urban drainage. *Urban Water Journal*. doi: 10.1080/1573062X.2014.916314.

Fowler D, Coyle M, Skiba U, et al. (2013) The global nitrogen cycle in the twenty-first century. *Philosophical Transactions of the Royal Society of London B*, 368, 1621. doi: 10.1098/rstb.2013.0165.

Galloway J N, Denterner F J, Capone D G, et al. (2004) Nitrogen cycles: past, present, future. *Biogeochemistry*, 70, 153-226.

George K, Ziska L H, Bunce J A, et al. (2007) Elevated atmospheric CO_2 concentration and temperature across an urban-rural transect. *Atmospheric Environment*, 41, 7654-7665.

Golubiewski N E. (2003) *Carbon in Conurbations: Afforestation and Carbon Storage as Consequences of Urban Sprawl in Colorado's Front Range*. PhD Thesis, University of Colorado at Boulder.

Grimm N B, Faeth S H, Golubiewski N E, et al. (2008a) Global change and the ecology of cities. *Science*, 319, 756-760.

Grimm N B, Foster D, Groffman P, et al. (2008b) The changing landscape: ecosystem responses to urbanization and pollution across climate and societal gradients. *Frontiers in Ecology and the Environment*, 6, 264-272.

Grimm N B, Redman C L. (2004) Approaches to the study of urban ecosystems: the case of Central Arizona-Phoenix. *Urban Ecosystems*, 7, 199-213.

Groffman P M, Law N L, Belt K T, et al. (2004) Nitrogen fluxes and retention in urban water-shed ecosystems. *Ecosystems*, 7, 393-403.

Gruber N, Galloway J N. (2008) An Earth-system perspective of the global nitrogen cycle. *Nature*, 451, 293-296.

Grubler A. (1994) Technology. In *Changes in Land Use and Land Cover: A Global Perspective*. William B M, Turner B L eds. Cambridge University Press, Cambridge, 287-328.

Gu B, Ge Y, Ren Y, et al. (2012) Atmospheric reactive nitrogen in China: sources, recent trends, damage costs. *Environmental Science and Technology*, 46, 9420-9427.

Gu B, Leach A M, Ma L, et al. (2013) Nitrogen footprint in China: food, energy, nonfood goods. *Environmental Science and Technology*, 47, 9217-9224.

Hammond G P, Jones C I. (2008) Embodied energy and carbon in construction materials. *Proceedings of the Institute of Civil Engineers-Energy*, 161, 87-98.

Handia L, Tembo J, Mwiinda C. (2004) Applicability of rainwater harvesting in urban Zambia. *Journal of Science and Technology*, 1, 1-8.

Hansell D A, Carlson C A, Repeta D J, et al. (2009) Dissolved organic matter in the ocean: a controversy stimulates new insights. *Oceanography*, 22, 202-211.

Hatt B E, Fletcher T D, Walsh C J, et al. (2004) The influence of urban density and drainage infrastructure on the concentration and loads of pollutants in small streams. *Environmental Management*, 34, 1112-1124.

Hill S. (2013) *Reforms for a Cleaner, Healthier Environment in China*. OECD Economics Department Working Papers 1045, OECD Publishing. Paris. 10.1787/5k480c2dh6kf-en(accessed 02/07/2015).

Hilten R N, Lawrence T M, Tollner E W. (2008) Modeling stormwater runoff from green roofs with HYDRUS-1D. *Journal of Hydrology*, 358, 288-293.

Hoegh-Guldberg O, Bruno J F. (2010) The impact of climate change on the world's marine ecosystems. *Science*, 328, 1523-1528.

Hoornweg D, Sugar L, Gomez C K T. (2011) Cities and greenhouse gas emissions: moving forward. *Environment and Urbanization*, 23, 207-227.

Hsieh C, Davis A P, Needelman B A. (2007) Nitrogen removal from urban stormwater runoff through layered bioretention columns. *Water Environment Research*, 79, 2404-2411.

Idso C D, Idso S B, Balling R C Jr, (1998) The urban CO_2 dome of Phoenix, Arizona. *Physical Geography*, 19, 95-108.

Idso C D, Idso S B, Balling R C Jr, (2001) An intensive two-week study of an urban CO_2 dome in Phoenix, Arizona, USA. *Atmospheric Environment*, 35, 995-1000.

Imberger S J, Thompson R M, Grace M R. (2011) Urban catchment hydrology overwhelms reach scale effects of riparian vegetation on organic matter dynamics. *Freshwater Biology*, 56, 1370-1389.

Imhoff M L, Bounoua L, Defries R, et al. (2004) The consequences of urban land transformation on net primary productivity in the United States. *Remote Sensing of Environment*, 89, 434-443.

IPCC. 2014. *Climate Change* 2014: *Synthesis Report. Contribution of Working Groups I, II and III to the Fifth Assessment Report of the Intergovernmental Panel on Climate Change*. Pachauri R K, Meyer L A eds. IPCC, Geneva.

Jagals P. (1996) An evaluation of sorbitol-fermenting bifidobacteria as specific indicators of human faecal pollution of environmental water. *South African Water Research Commission*, 22, 235.

Jo H, McPherson G. (1995) Carbon storage and flux in urban residential greenspace. *Journal of Environmental Management*, 45, 109-133.

Joos F, Roth R, Fuglestvedt J S, et al. (2013) Carbon dioxide and climate impulse response functions for the computation of greenhouse gas metrics: a multi-model analysis. *Atmospheric Chemistry and Physics*, 13, 2793-2825.

Kajikawa A, Inoue T, Goh T N. (2011) Analysis of building environment assessment frameworks and their implications for sustainability indicators. *Sustainability Science*, 6, 233-246.

Katsiri A. (2009) Access to improved sanitation and wastewater treatment. *ENHIS-European Environmental and Health Information Systems*. WHO Regional Office for Europe, Copenhagen. http://www.euro.who.int/__data/assets/pdf_file/0009/96957/1.3.-Access-to-improved-sanitation-and-wastewater-treatment-EDITED_layouted.pdf?ua=1(accessed 02/07/2015).

Kaye J P, Groffman P M, Grimm N B, et al. (2006) A distinct urban biogeochemistry? *Trends in Ecology and Evolution*, 21, 192-199.

Keith D A, Rodriguez J P, Rodriguez-Clark K M, et al. (2013) Scientific foundations for an IUCN Red List of ecosystems. *PLoS ONE* 8, e62111.

Kennedy C, Cuddihy J, Engel-Yan J. (2007) The changing metabolism of cities. *Journal of Industrial Ecology*, 11, 43-59.

Kennedy C, Steinberger J, Gasson B, et al. (2009) Greenhouse gas emissions from global cities. *Environmental Science and Technology*, 43, 7297-7302.

Kissinger M. (2012) Internationaltrade related food miles—the case of Canada. *Food Policy*, 37, 171-178.

Kondoh A, Nishiyama J. (2000) Changes in hydrological cycle dueto urbanization in the suburb of Tokyo Metropolitan Area, Japan. *Advances in Space Research*, 26, 1173-1176.

Koont S. (2011) *Sustainable Urban Agriculture in Cuba*. University Press of Florida, Florida.

Lee J H, Bang K W, Ketchum L H, et al. (2002) First flush analysis of urban storm runoff. *The Science of the Total Environment*, 293, 163-175.

Leopold L B. (1968) *Hydrology for Urban Land Planning: a Guidebook on the Hydrological Effects of Urban Land Use*. US Geological Survey Circular 554. USGS, Washington DC. http://pubs.usgs.gov/circ/1968/0554/report.pdf(accessed 02/07/2015).

Le Quéré C, Moriarty R, Andrew R M, et al. (2015) Global Carbon Budget 2014. *Earth System Science Data*. doi: 10.5194/essd-7-47-2015 http://www.earth-syst-sci-data.net/7/47/2015/essd-7-47-2015.html(accessed 02/07/2015).

Li L, Yin C, He Q, et al. (2007) First flush of storm runoff pollution from an urban catchment in China. *Journal of Environmental Sciences*, 19, 295-299.

Likens G E. (1992) *The Ecosystem Approach: Its Use and Abuse.* Excellence in Ecology, Vol. 3. Ecology Institute, Oldendorf/Luhe.

Lindh G. (1972) Urbanization: a hydrological headache. *Ambio*, 1, 185-201.

Loreau M. (2010) Linking biodiversity and ecosystems: towards a unifying ecological theory. *Philosophical Transactions of the Royal Society B*, 365, 49-60.

Lovell S T. (2010) Multifunctional urban agriculture for sustainable land use planning in the United States. *Sustainability*, 2, 2499-2522.

Lu C, Tian H. (2007) Spatial and temporal patterns of nitrogen deposition in China: Synthesis of observational data. *Journal of Geophysical Research* 112, D22S05. doi: 10.1029/2006JD007990.

Luck G W, Daily G C, Ehrlich P R. (2003) Population diversity and ecosystem services. *Trends in Ecology and Evolution*, 18, 331-336.

Mahasenan N, Smith S, Humphreys K. (2003) The cement industry and global climate change: current and potential future cement industry CO_2 emissions. Gale J, Kaya Y eds. *Greenhouse Gas Control Technologies-6th International Conference.* Pergamon, Oxford, 995-1000. 10.1016/B978-008044276-1/50157-4(accessed 02/07/2015).

Manzoni S, Trofymow J A, Jackson R B, et al. (2010) Stoichiometric controls on carbon, nitrogen, phosphorus dynamics in decomposing litter. *Ecological Monographs*, 80, 89-106.

Maurin F D. (2011) Fukushima: Consequences of systemic problems in nuclear plant design. *Economic and Political Weekly*, 46, 10-12.

McClintock N. (2010) Why farm a city? Theorizing urban agriculture through a lens of metabolic rift. *Cambridge Journal of Regions, Economy and Society*, 3, 191-207.

McPherson G, Simpson J R, Peper P J, et al. (2005) Municipal forest benefits and costs in five US cities. *Journal of Forestry*, 103, 411-416.

Mentens J, Raes D, Hermy M. (2006) Green roofs as a tool for solving the rain-water runoff problem in the urbanized 21st century? *Landscape and Urban Planning*, 77,

217-226.

Millennium Ecosystem Assessment (2003) *Ecosystems and Human Well-Being: A Framework for Assessment*, Island Press, Washington DC.

Mok H, Williamson V G, Grove J R, et al. (2014) Strawberry fields forever? Urban agriculture in developed countries: a review. *Agronomy for Sustainable Development*, 34, 21-43.

Moomaw W R. (2002) Energy, industry and nitrogen: strategies for decreasing reactive nitrogen emissions. *Ambio*, 31, 184-189.

Moomaw W R, Burgherr P, Heath G, et al. (2011) Annex II: Methodology. In *IPCC Special Report on Renewable Energy Sources and Climate Change Mitigation*. Edenhofer O, Pich-Madruga R, Sokona Y, et al. eds. Cambridge University Press, Cambridge, 973-1000.

Mougeot L J A. (2005) *AGROPOLIS: The Social, Political, Environmental Dimensions of Urban Agriculture*, Earthscan, London.

Mu Q, Zhao M, Running S W. (2011) Improvements to a MODIS global terrestrial evapotranspiration algorithm. *Remote Sensing of Environment*, 115, 1781-1800.

Norton B A, Bosomworth K, Coutts A, et al. (2013) *Planning for a Cooler Future: Green Infrastructure to Reduce Urban Heat*. Victorian Centre for Climate Change Adaptation Research, Melbourne. http://www.vcccar.org.au/sites/default/files/publications/VCCCAR Green Infrastructure Guide Final.pdf (accessed 25/06/2015).

Norton B A, Coutts A M, Livesley S J, et al. (2015) Planning for cooler cities: A framework to prioritise green infrastructure to mitigate high temperatures in urban landscapes. *Landscape and Urban Planning*, 134, 127-138.

Nowak D J, Crane D E. (2002) Carbon storage and sequestration by urban trees in the USA. *Environmental Pollution*, 116, 381-389.

Nowak D J, Crane D E, Stevens J C. (2006) Air pollution removal by urban trees and shrubs in the United States. *Urban Forestry and Urban Greening*, 4, 115-123.

Oberndorfer E, Lundholm J, Bass B, et al. (2007) Green roofs as urban ecosystems:

ecological structures, functions, services. *BioScience*, 57, 823-833.

O'Brien D, Hes D. (2008) The third way: developing low environmental impact housing prototypes for hot/humid climates. *International Journal for Housing Science*, 32, 311-322.

Odum E P. (1983) *Basic Ecology*. Saunders Publishing, Philadelphia.

Okoko E. (2008) The urban storm water crisis and the way out: empirical evidences from Ondo Town, Nigeria. *The Social Sciences*, 3, 148-156.

Olivier J G J, Janssens-Maenhout G, Peters J A H W, et al. (2011) *Long-term Trend in Global CO_2 Emissions*. 2011 Report. PBL Netherlands Environmental Assessment Agency and theEuropean Union, The Hague. http://www.pbl.nl/sites/default/files/cms/publicaties/C02 Mondiaal_ webdef_19sept. pdf(accessed 03/07/2015).

Owusu-Asante Y, Ndiritu J. (2009) The simple modelling method for storm- and grey-water quality management applied to Alexandra settlement. *Water SA*, 35, 615-626.

Owusu-Asante Y, Stephenson D. (2006) Estimation of storm runoff loads based on rainfall-related variables and power law models-case study in Alexandra. *Water SA*, 32, 1-8.

Parkinson J. (2003) Drainage and stormwater management strategies for low-income urban communities. *Environment and Urbanization*, 15, 115-126.

Parkinson J, Tayler K, Mark O. (2007) Planning and design of urban drainage systems in informal settlements in developing countries. *Urban Water Journal*, 4, 137-149.

Parris K M, Hazell D L. (2005) Biotic effects of climate change in urban environments: the case of the grey-headed flying-fox (*Pteropus poliocephalus*) in Melbourne. Australia. *Biological Conservation*., 124, 267-276.

Pataki D E, Alig R J, Fung A S, et al. (2006) Urban ecosystems and the North American carbon cycle. *Global Change Biology*, 12, 2092-2102.

Pauchard A, Aguayo M, Pena E, et al. (2006) Multiple effects of urbanization on the biodiversity of developing countries: the case of a fast-growing metropolitan area (Concepción, Chile). *Biological Conservation*, 127, 272-281.

Paul M J, Meyer J L. (2001) Streams in the urban landscape. *Annual Review of Ecology* and Systematics, 32, 333-365.

Pickett S T A, Cadenasso M L, Grove J M, et al. (2001) Urban ecological systems: linking terrestrial ecological, physical, socioeconomic components of metropolitan areas. *Annual Review of Ecology, Evolution, Systematics*, 32, 127-157.

Pickett S T A, Grove J M. (2009) Urban ecosystems: what would Tansley do? *Urban Ecosystems*, 12, 1-8.

Pidwirny M. (2008) *Fundamentals of Physical Geography*, 2nd ed (online text book). Physical Geography. net.

Pouyat R V, Russell-Anell J, Yesilonis I D, et al. (2003) *Soil Carbon in Urban Forest Ecosystems*. CRC Press, Boca Raton.

Prather M J, Holmes C D, Hsu J. (2012) Reactive greenhouse gas scenarios: systematic exploration of uncertainties and the role of atmospheric chemistry. *Geophysical Research Letters* 39, L09803. doi: 10.1029/2012GL051440.

Prentice I C, Farquhar G D, Fasham M J R, et al. (2001) The carbon cycle and atmospheric carbon dioxide. In *Climate Change* 2001: *The Scientific Basis*. Houghton J T, Ding Y, Griggs D Y, et al. eds. Cambridge University Press, Cambridge, 183-237.

Rabalais N N. (2002) Nitrogen in aquatic ecosystems. *AMBIO: A Journal of the Human Environment*, 31, 102-112.

Reddy B V V, Jagadish K S. (2003) Embodied energy of common and alternative building materials and technologies. *Energy and Buildings*, 35, 129-137.

Reich P B, Hobbie S E. (2013) Decade-long soil nitrogen constraints on the CO_2 fertilization of plant biomass. *Nature Climate Change*, 3, 278-282.

Roth M. (2007) Review of urban climate research in (sub) tropical regions. *International Journal of Climatology*, 27, 1859-1873.

Seitzinger S P, Kroeze C. (1998) Global distribution of nitrous oxide production and N inputs in freshwater and coastal marine ecosystems. *Global Biogeochemical Cycles*, 12, 93-113.

Smit J, Nasr J. (1992) Urban agriculture for sustainable cities: using wastes and idle land and water bodies as resources. *Environment and Urbanization*, 4, 141-152.

Stoffberg G H, van Rooyen M W, van der Linde M J, et al. (2010) Carbon sequestration estimates of indigenous street trees in the city of Tshwane, South Africa. *Urban Forestry and Urban Greening*, 9, 9-14.

Stone B Jr. (2004) Paving over paradise: how land use regulations promote residential imperviousness. *Landscape and Urban Planning*, 69, 101-113.

Suppiah R, Whetton P H. (2007) *Projected Changes in Temperature and Heating Degree-Days for Melbourne and Victoria*, 2008-2012. CSIRO Marine and Atmospheric Research, Aspendale. http://www.ccma.vic.gov.au/soilhealth/climate_change_literature_review/documents/organisations/csiro/MelbourneEDD2008_2012.pdf (accessed 02/07/2015).

Svirejeva-Hopkins A, Schellnhuber H J, Pomaz V L. (2004) Urbanised territories as a specific component of the global carbon cycle. *Ecological Modelling*, 173, 295-312.

Tansley A G. (1935) The use and abuse of vegetation concepts and terms. *Ecology*, 16, 284-307.

Tarnocai C, Canadell J G, Schuur E A G, et al. (2009) Soil organic carbon pools in the northern circumpolar permafrost region. Global Biogeochemical Cycles 23, GB2023. DOI: 10.1029/2008GB003327.

Tarvainen O, Markkola A M, Strommer R. (2003) Diversity of macrofungi and plants in Scots pine forests along an urban pollution gradient. *Basic and Applied Ecology*, 4, 547-556.

Taylor S L, Roberts S C, Walsh C J, et al. (2004) Catchment urbanization and increased benthic algal biomass in streams: linking mechanisms to management. *Freshwater Biology*, 49, 835-851.

Tscharntke T, Klein A M, Steffan-Dewenter I, et al. (2005) Landscape perspectives on agriculture intensification and biodiversity-ecosystem service management. *Ecology Letters*, 8, 857-874.

Turner M G, Collins S L, Lugo A L, et al. (2003) Disturbance dynamics and ecological response: the contribution of long-term ecological research. *BioScience*,

53, 46-56.

Tyrvainen L, Pauleit S, Seeland K, et al. (2005) Benefits and uses of urban forests and trees. In *Urban Forests and Trees*. Konijnendijk C C, Nilsson K, Randrup T B, et al. eds. Springer, Berlin Heidelberg, 81-114.

United Nations. (1998) *Kyoto Protocol to the United Nations Framework Convention on Climate Change*. United Nations Framework Convention on Climate Change, Bonn. http://unfccc.int/resource/docs/convkp/kpeng.pdf (accessed 02/07/2015).

United Nations. (2012) *Doha Amendment to the Kyoto Protocol*. United Nations Framework Convention on Climate Change, Bonn. http://unfccc.int/files/kyoto_protocol/application/pdf/kp_doha_amendment_english.pdf (accessed 02/07/2015).

United Nations. (2014) *World Urbanization Prospects: The* 2014 *Revision, CD-ROM Edition*. United Nations Department of Economic and Social Affairs, Population Division, New York. http://esa.un.org/unpd/wup/ (accessed 08/06/2015).

United Nations. (2015) *Adoption of the Paris Agreement*. United Nations Framework Convention on Climate Change, Bonn. http://unfccc.int/resource/docs/2015/cop21/eng/l09r01.pdf (accessed 21/01/2016).

van Aardenne J A, Dentener F J, Olivier J G J, et al. (2001) A 1×1 resolution data set of historical anthropogenic trace gas emissions for the period 1890-1990. *Global Biogeochemical Cycles*, 15, 909-928.

van Veenhuizen R. (2006) *Cities Farming for the Future: Urban Agriculture for Green and Productive Cities*. RUAF Foundation, IDRC and IIRR, Silang.

Vitousek P M, Aber J D, Howarth R W, et al. (1997) Human alteration of the global nitrogen cycle: Sources and consequences. *Ecological Applications*, 7, 737-750.

Walsh C J, Fletcher T D, Burns M J. (2012) Urban stormwater runoff: a new class of environmental flow problem. *PLoS ONE*, 7, 1-10.

Walsh C J, Roy A H, Feminella J W, et al. (2005) The urban stream syndrome: current knowledge and the search for a cure. *Journal of North American Benthological*

Society, 24, 706-723.

Wang D, Heckathorn S A, Wang X, et al. (2012) A meta-analysis of plant physiological and growth responses to temperature and elevated CO_2. *Oecologia*, 169, 1-13.

White P S, Pickett S T A. (1985) The ecology of natural disturbance and patch dynamics. In *The Ecology of Natural Disturbance and Patch Dynamics*. Pickett S T A, White P S eds. Academic Press, New York, 3-13.

Willis A J. (1997) Forum. *Functional Ecology*, 11, 268-271.

Wong T H F. (2007) Water sensitive urban design: the journey thus far. *Australian Journal of Water Resources*, 110, 213-222.

World Steel Association. (2013) *Steel's Contribution to a Low Carbon Future: World Steel position paper*. World Steel Association, Brussels. http://www.worldsteel.org/publications/position-papers/Steels-contribution-to-a-low-carbon-future.html (accessed 06/06/2015).

World Steel Association. (2014) *Crude steel production* 2014-2015. World Steel Association, Brussels. http://www.worldsteel.org/statistics/crude-steel-production.html (accessed 02/06/2015).

Yudelson J. (2007) *Green Building A to Z: Understanding the Language of Green Building*. New Society Publishers, Gabriola Island.

Zehr J P, Kudela R M. (2011) Nitrogen cycle of the open ocean: from genes to ecosystems. *Annual Review of Marine Science*, 3, 197-225.

第 6 章 人类的城市生态学

6.1 概述

在这个加速城镇化的时代,城市人口数量比以往任何时候都多——无论是绝对数量还是占总人口的比例。在最好的情况下,城市是人类事业、创造力、社区和想要生活的中心,为大量人口带来地方感、多样性和归属感。城市提供机会、就业、参与、成就感和对更美好未来的憧憬(Hes, Du Plessis, 2015)。一般而言,城市地区的贫困率低于乡村地区(世界银行, 2013; 图 6.1)。设计良好的城市,提供了一种节省空间和能源的方式,可以大规模地满足人类经济需求和社会需求(Kennedy et al., 2007; Kennedy et al., 2011; Farr, 2012)。城市中最富裕的居民享有高水平的生活和舒适的设施,包括安全的房屋、电力,清洁的水,充足的通风,有效的供暖和降温,以及获得优质教育和医疗保健。富裕的城市社区往往整齐有序,公园和街道树木丰富,在夏季可以提供舒适、阴凉的微气候环境(Heynen, 2003; Harlan et al., 2006; Zhu, Zhang, 2008; Huang et al., 2011b; 图 6.2(a))。

但是,在最坏的情况下,城市可能成为失业、暴力、贫困、污垢和人类绝望的中心。城市广阔、人口密集以及混凝土和沥青中的建筑灰色,可能使城市变得缺乏人情味和疏远。许多城市中最贫穷的居民,生活在缺乏正规、人满为患、没有安全保障的住房中,常常缺少有效的供暖和降温,充足的通风和卫生设施,或无法获得干净的饮用水(Weeks et al., 2007; Montgomery, 2009)。在大多数城市中,贫困社区的植被覆盖率相对较低,公园和街道树木较少,不透水表面覆盖率较高(Pedlowski et al., 2002; Heynen, 2003; Escobedo, 2006; Harlan et al.,

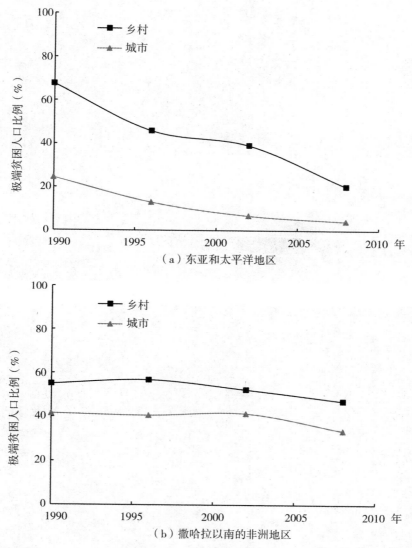

图 6.1 1990—2008 年,在世界两个区域中生活在极端贫困中的乡村和城市人口百分比:(a) 东亚和太平洋;(b)撒哈拉以南非洲。极端贫困的定义是每天生活费低于 1.25 美元(世界银行数据,2013)

2006; Weeks et al. ,2007; Huang et al. ,2011b; Shanahan et al. ,2011; Clarke et al. ,2013;图 6.2(b))。例如,非洲加纳的贫民窟的植被覆盖率估计为 6%,不

(a)

(b)

图 6.2 城市绿地覆盖率对比：(a) 美国佐治亚亚特兰大的植物园。亚特兰大被称为"树木之城"，在城市范围内有 48% 的林木覆盖率(Giarrusso et al., 2014)；(b) 巴西的一个非正规居住区(贫民窟)(拍摄：Unsplash)

透水覆盖率为84%(主要为密集的屋顶),裸露的土壤覆盖率为10%(Weeks et al.,2007)。即使是规划了的城市(非贫民窟),植被覆盖率也会随着邻里的社会经济状况而变化。例如,智利圣地亚哥的富裕城市和乡镇的树木平均覆盖率为33%,贫穷城市的树木平均覆盖率为12%(Escobedo et al.,2006)。

我们如何以最好的方式构思人类城市生态学?研究城市与社会的每门学科,包括社会学、人文地理学、城市规划、人口学、心理学、犯罪学、流行病学和公共卫生都有自己的理论基础和方法论。生态学也可以应用于人类。人类生态学是研究人类与自然、社会和人工环境之间的关系。它通常被概念化为涵盖地理、人类学、心理学、生物学、社会学和城市规划的跨学科领域(The Editors,1972;Bates,2012)。这些学科对了解城市和管理城市都很重要。但是,在本章中,我将从单一学科(生态学和生物学)的角度考虑城镇化进程和城市环境中的生活经验如何影响作为个体和种群的人类。城市环境的特征从城市形态的性质到空气污染、热量、噪声、清洁饮用水和卫生设施的获取,这些对每个人都至关重要(Schell,Denham,2003;Miller et al.,2007;Valavanidis et al.,2008;Montgomery,2009;Hammer et al.,2014)。从个体到种群,这些特征还影响世界各地城市人口的关键比率(出生、死亡、迁入和迁出)。

该分析的核心是人类城市体验的空间多样性,无论是在单个城市内部以及不同城市之间。随着时间的推移,人类城市体验也会有所不同,因为城市从其早期基础到扩张和工业化,随后是衰退,城市没落,以及在某些情况下进一步的城市更新,都会遵循不同的轨迹(Wyly,Hammel,1999;Martinez-Fernandez et al.,2012;Zheng et al.,2014)。《电报路》(*Telegraph Road*)的歌词中捕捉到这样一种轨迹(见"序二"),它描述了密歇根底特律的一条通道从土路到多车道的转变;它周围的城市从单一的小木屋变成工业化大都市;居民经历了从早期定居、城市扩张、经济繁荣时期的乐观主义,到失业、城市衰退和疏远的绝望(Knopfler,1982)。目前,非洲、亚洲和南美洲的城市快速扩张,在很大程度上是由于贫困地区乡村人口大规模移民造成的,而且在许多情况下,其特征是建造了非正规的居住区,如贫民窟和棚户区(UN-Habitat,2003;Cilliers,2008;WHO,UN-Habitat,2010;Shin,Li,2012)。这些新移民的城市体验将与长期居住在城市规划良好且资源相对充足的地区的居民完全不同。

6.2 城市形态

6.2.1 城市公园和开敞空间

正如前面各章节所讨论的，城市建设涉及多种过程，这些过程将景观的绿色和棕色、软质表面和开敞空间，清洁的空气和水，自然的声音和夜间的黑暗变为灰色、坚硬的表面，密集的建筑物和道路，受污染的空气和水，以及人为噪声和夜间照明。城市的这些特征影响着居住在这里的人的健康和福祉，使他们与自然分离，并在某些情况下使他们彼此分离（Silver，1997；Turner et al.，2004；Miller，2005；Soule，2009；Ross，2011；见专栏6.1）。许多城市居民居住在生物条件恶劣的社区中，因此只能与数量稀少的非人类物种互动。

> **专栏6.1**
>
> **城市中人-野生生物互动**
>
> 在私人花园、街道景观或公园和保护区中与野生动物（即非驯化动物）互动，为城市居民提供了更多在城市景观中与自然联系的机会。从简单观察无脊椎动物、鸟类、爬行动物、青蛙和其他动物，到提供食物和庇护所、争夺资源、破坏财产、传播疾病和(人类和野生动植物)伤害或死亡，人类和野生动物之间的相互作用可以按照强度梯度进行分类。城市中，人类与野生动植物的积极互动可以是互利的。例如，许多人喜欢在自己的花园或附近区域观察鸟类，而鸟类浴池和喂鸟器可以为鸟类提供重要的资源（Miller，2005；Jones，Reynolds，2008；图6.3）。尽管城市环境的特征发生了很大变化，但是饲喂花园中的鸟仍被视为人与自然之间持续联系的一种物理表现（Fuller at al. 2008），而且这种联系对于人类的福祉至关重要（Maller，Townsend，2004；Miller，2005；Jones，Reynolds，2008）。
>
> 鸟类的补充喂食增加了单个花园的鸟类多样性和多度，以及整个社区的鸟类总多度（Chamberlain et al.，2005；Daniels，Kirkpatrick，2006；Parsons

et al., 2006; Fuller et al., 2008)。然而，某些食物类型的大量供应(例如，英国估计每年供应6万吨种子和花生；Glue, 2006)，可以更有利于食谷鸟类，而不是食虫和食蜜鸟类，这会改变种群的组成(Allen, O'Connor, 2000)。另外，给鸭子和其他水禽喂食面包，可能会增加城市鸟类群落的种内和种间攻击，并增加对营养质量差的食物来源的依赖(Campbell, 2008; Chapman, Jones, 2009; Chapman, Jones, 2011)。

在世界许多地方，随着城市扩展到以前的野生栖息地，城市环境中人类与野生动物之间的直接和间接冲突正在增加。当动物违背人类意愿利用城市中特定食物资源，例如私人花园里的鲜花、水果和蔬菜，食物残渣、垃圾，以及人类为宠物或另一类野生动物(例如鸟类)提供的食物时，会产生冲突(Banks et al., 2003; Beckmann, Berger, 2003; Harper et al., 2008; Lamarque et al., 2009; Adams et al., 2013)。某些中型捕食者，例如赤狐、浣熊、欧亚獾，非常适应城市生活，并且在城市和郊区，种群密度高于乡村地区(Bateman, Fleming, 2012; Šálek et al., 2015)。在澳大利亚墨尔本，引入的赤狐通常会杀死家鸡以及本地野生动物，如濒危的澳棕短鼻袋狸、沼泽石龙子(小蜥蜴)、咆哮草蛙(*Litoria raniformis*)(Saunders et al., 1995; Green, 2014)。

较大的食肉动物，包括美洲狮、豹子、狮子，可能捕食诸如猫和狗之类的家庭宠物，在某些情况下还会捕食人(McKee, 2003; Thornton, Quinn, 2009; Bhatia et al., 2013)。在北美，土狼和美洲狮有时会攻击家畜和人类(Beier, 1991; Thornton, Quinn, 2009; White, Gehrt, 2009)。在美国和加拿大的城市和郊区环境中，土狼袭击的发生率似乎随着土狼数量的增长而增加(White, Gehrt, 2009)。在印度孟买，2000年至2010年，证实发生了101起豹子袭击人类事件，造成人的伤亡，大部分袭击发生在2001年至2005年之间(Bhatia et al., 2013)。这些豹子属于甘地国家公园的种群，该公园毗邻孟买的居民区，该地区的人口密度超过了3万/km²。在许多人与野生动物冲突的情况下，人类利益优先于野生动植物的利益，捕获并安置或安乐死"问题"动物(Choudhury, 2004; Spencer et al., 2007; Bhatia et al., 2013)。但是，为了避免人类与野生动物的冲突而将某些动物重新安置，这些安置的

图 6.3　在后院喂食器上的东玫瑰鹦鹉（*Platycercus eximius*）（拍摄：Lesley Smitheringale）

成功率很小，许多迁徙者返回了捕获地点，而另一种情况是这些动物在新环境中遭受了高死亡率（Pietsch，1994；Linnell et al.，1997；Sullivan et al.，2004；Germano，Bishop，2009）。

例如，在一项对五个城市（亚利桑那州图森；德国柏林；华盛顿特区；意大利佛罗伦萨；日本千叶市）的社区生物多样性研究中，多达84%的居民生活区的鸟类和蕨类植物的多样性低于平均水平（Turner et al.，2004）。这是由于人口密度和生物多样性之间存在负相关关系，而城市中生物多样性更多地（例如自然公园和残留植被的区域）聚集在城市边缘或其他单位面积人口较少的社区。因此，许多居民与城市仍然设法支持的大部分生物多样性分离（Miller，2005）。城市中，人与自然的隔阂会导致一种被称为"经验灭绝"的现象（Pyle，1978）；一个经验匮乏的循环，首先是当地动植物的丧失，然后是对自然的疏远和冷漠。反过来，这可能导致生物多样性的进一步丧失,以及与自然界的更深隔离(Pyle,1978;Miller,2005)。有人提出,这种隔离可能会减少人类对更广泛的生物保护的义务(即在城市范围以外)(Miller,2005),但它也会对人类健康和福祉有重大影响。

在城市保护区、社区公园,甚至通过一棵树与自然联系,可以对人的精神健康和心理健康产生一系列积极影响,减轻压力和精神疲劳,恢复注意力,增强思考和获得视野,并提供对地方的依恋感和更高的生活满意度(Kuo,2001;Grahn,Stigsdotter,2003;Chiesura,2004;Fuller et al.,2007;Chaudhry et al.,2011;Haq,2011;Maller et al.,2004;Clark et al.,2014;Shanahan et al.,2015)。例如,一项关于"绿色运动"(在自然环境中锻炼)对心理健康益处的元分析发现,在10项研究中,其对自尊和情绪具有一致的积极影响[总体影响大小:自尊的$d=0.46$,CI = 0.34~0.59,情绪的 $d=0.54$,CI = 0.38~0.69;Barton,Pretty,2010]。有趣的是,量效关系表明,仅进行五分钟的绿色锻炼,即可获得实质性的心理健康益处,并且在城市绿色空间中锻炼,对情绪和自尊的积极影响与在某些乡村绿地中(如农田、森林和林地)锻炼相类似。但是,在绿色自然环境中锻炼并欣赏水景,可以为情绪和自尊带来更大的益处。

公园和共用花园也可以作为社区参与的焦点,增加邻居之间的互动,提高社会凝聚力和加强社会纽带(Coley et al.,1997;Saldivar-Tanaka,Krasny,2004;Kingsley,Townsend,2006;Maas et al.,2009)。一项对荷兰1万名居民的调查发现,一个人的家周围1km和3km半径范围内,绿地覆盖率越高,人的孤独感和社会孤立感就越低,这表明更多的社会接触,可能是绿色空间和健康之间正相关的机制之一(Maas et al.,2009)。简单地通过窗户观看绿色的屋顶花园,可以改善市政工作人员的注意力和心情(Lee et al.,2015),而一项经典研究发现(Ulrich,1984),如果患者从医院的窗户可以看到自然风景,则可以更快地从手术中康复。有趣的是,在英国谢菲尔德的一项研究发现,人们从城市绿色空间中获得的心理幸福感随着人感知到的物种丰富度的增加而增加,这表明生物复杂性是人类享受大自然的关键组成部分,因此是城市公园的宝贵特征(Fuller et al.,2007;Carrus et al.,2015)。

城市开敞空间通过提供运动和娱乐的空间和设施,改善了城市居民的身体健康状况(Sallis et al.,2011;Lowe et al.,2014)。特别是在一些富裕国家,运动不足易导致患肥胖、心脏病、癌症和 2 型糖尿病等慢性病（Kaczynski, Henderson，2007；Bauman et al.，2012；Lowe et al.，2014）。慢性病已经超过传染病（专栏 6.2），成为城市人口患病和死亡的主要原因（Lowe et al.，2014）。来自全球 122 个国家/地区的最新数据表明，平均而言，31.1%（95%，CI＝30.9%～31.2%）的成年人和 80.3%（80.1%～80.5%）的青少年未达到建议的运动水平（Hallal et al.，2012）。不做运动水平因地区而异，从东南亚的 17%（16.8%～17.2%）到美洲、东地中海的大于 40%（图 6.4）。城市居民区公园的可用性、邻近性、可达性和公园质量，都与居民中较高的体育活动有关（Veitch et al.，2014）。公园可以为人们提供散步或骑自行车的适宜环境，从而鼓励人们锻炼（Pikora et al.，2003；Cervero et al.，2009；Sugiyama et al.，2010，2012）。在澳大利亚珀斯，与交通条件差的居民相比，拥有便利到达的大而有吸引力的公共开敞空间的居民，经常散步的可能性要高出 50%（每周散步六次或以上，总计大于或等于 3 小时）（Giles-Corti et al.，2005）。

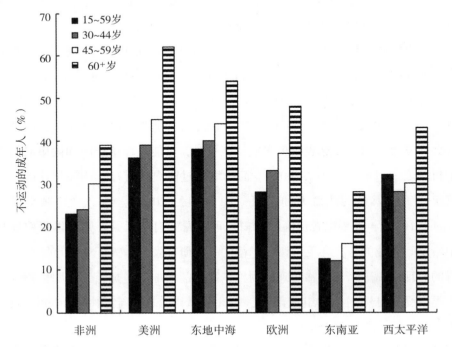

图 6.4 按年龄组别和世界卫生组织区域划分的缺乏体育活动情况（占人口百分比）；运动不足因地区而异，并且会随着年龄的增长而增加（Hallal et al.，2012；图 2）

> 专栏6.2

城市环境中的传染病

纵观整个历史,城市中人口密度高、营养不良和恶劣的卫生条件共同促进了传染病的传播(Redman,Jones,2004)。世界上许多早期的城市促成了一系列传染病的演变和传播,这些传染基本已导致数百年来广泛的发病率和死亡率(Redman,1999)。这种趋势早在工业革命期间仍然存在,当时西欧和英国拥挤不堪的城市是诸如天花、痢疾、肺结核、伤寒、霍乱等疾病的中心。霍乱是摄入被感染者的粪便污染的水或食物所传播的细菌引起的。当饮用水供应被污水或未经处理的人类废物污染时,该疾病会迅速蔓延。

霍乱的暴发在19世纪伦敦修建城市下水道之前是很普遍的,最著名的暴发发生在1854年的索霍区布罗德街。当地医生John Snow绘制了霍乱病例的分布图,并确定了布罗德街的一个水泵是感染源,因为几乎所有死亡都发生在该泵很近的地方(图6.5;Johnson,2006)。他说服当地禁用此泵,即使当时疾病的细菌理论尚未被广泛接受。该事件被认为是流行病学发展的基础时刻。霍乱仍然在城市中发生,如2008年津巴布韦哈维拉雷暴发的霍乱,这是由于城市供水和卫生系统的崩溃以及雨季的来临造成的(WHO,2011)。这次暴发遍及津巴布韦,并扩散到许多邻国,估计死亡人数大于4000(Mason,2009;Mukandavire et al.,2011)。今天,在城市(尤其是非正规居住区)中常见的其他传染病包括结核病、登革热、艾滋病、肺炎和腹泻(WHO,UN-Habitat,2010)。

6.2.2 城市扩张和汽车依赖

低城市密度(也称为城市扩张)增加了对私家车的依赖,同时减少了居民在社区附近步行以及骑自行车上下班的机会(Ewing et al.,2003;Dieleman,Wegener,2004)。远离工作场所、公共交通以及商店、学校等便利设施的庞大住宅开发项目,是世界许多城市的特征。这些类型的社区不利于体育锻炼,并且与

图6.5 由英国医师、医学流行病学的创始人之一 John Snow 博士绘制的原始地图的变形版，显示1854年伦敦索霍区的霍乱死亡地点(黑色圆圈)。感染源是布罗德街的一台水泵

居民中的超重和肥胖呈正相关(Lopez, 2004; Garden, Jalaludin, 2009)。一项对佐治亚州亚特兰大市城市居民的研究发现，每个家庭1km以内的土地利用组合（通过四种土地利用类型的覆盖率计算：住宅、商业、办公和公共机构），与所有性别和族裔人群的肥胖概率有关($BMI > 30kg/m^2$)(Frank et al., 2004；图6.6)。每天增加1小时的汽车行程，肥胖的概率就会增加6%，而每天步行1km，肥胖的概率就会降低4.8%(Frank et al., 2004)。

6.2 城市形态

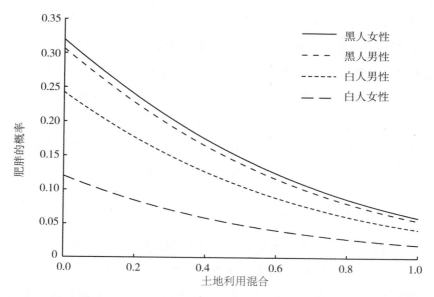

图6.6 在美国佐治亚州亚特兰大市，肥胖的概率与城市土地利用结构、种族和性别的关系。随着土地利用组合的增加，所有社会群体中肥胖的可能性都会下降。肥胖定义为体重系数 BMI≥30。土地利用混合是指每个参与者家庭 1km 以内的四种土地利用类型的覆盖均匀度（住宅、商业、办公和公共机构）(Frank et al., 2004, 图2)

较高的居住密度，良好的街道连通性（网格化的街道网络，而不是许多不相连的街道和小路），步行和自行车道的使用以及混合土地利用（住宅、商店和工作场所之间的距离更短），与较高的步行和骑自行车出行相关（Lowe et al., 2014）。近年来，城市社区的可步行性得到越来越多的研究和政策关注（WAPC, 2007; NSW Government, 2004; Saelens et al., 2003; Giles-Corti et al., 2015），并且社区的步行能力与其居民之间的肥胖率呈负相关（Smith, 2008; Sallis et al., 2009; Casagrande et al., 2011）。可以扩大这些个体水平的影响，并将其应用于研究肥胖和慢性疾病死亡率之间的关系，以估计城市扩张或社区可步行性对人口死亡率的影响（Hankey et al., 2012）。

可以通过三个"D"来识别适合步行的社区：人口密度（Population Density）、友好的人行道设计（Pedestrian-friendly Design）和目的地多样性（Diversity of Destination）(Cervero, Kockelman, 1997; Smith, 2008)。例如，道路的密度和连

185

通性在很大程度上影响着哥伦比亚波哥大的步行和骑行的比率(Cervero et al.，2009)。分析以居住区周围500m为半径的街区，在连通性中等或较高的社区中，居民每天步行≥30分钟的概率是连通性较低的社区的两倍。同样，非娱乐目的中，如果居住区半径1km以内的道路密度≥0.2km/km^2，居民因非娱乐目的(用于工作、上学、购物和看医生的出行)骑自行车≥30分钟/工作日的概率是道路密度<0.2km/km^{2}的居住区的居民的两倍(Cervero et al.，2009)。在许多城市中，较老的街区更适合行人，有阴凉的步行道、狭窄的街道、较慢的交通和更多样化的目的地；较新的社区更可能被设计成有效的交通流(Smith，2008)。

6.2.3 社区劣势和社区失调

城市形态的特征不仅促进或阻碍了城市居民的体育活动以及其与公园和保护区的自然联系，而且还反映了社区的社会秩序水平。社区失调是一个用于描述对人身安全有严重威胁的地方的术语(Ross，2011)。社区失调反映了社会的混乱——社会过程和结构的瓦解，而这些过程和结构在社区中的作用是维护尊重和安全(Kim，2008)。社区失调的迹象包括破旧的建筑物，用木板封住的窗户，涂鸦、噪声、垃圾、犯罪、在公共场所酗酒、公开销售和使用非法药物，以及青少年犯罪和街头帮派的存在(Galea et al.，2005；Kim，2008；Ross，Mirowsky，2009)。在一个相关的概念中，社区劣势描述的是缺乏经济和社会资源的地方(Ross，2011)。处境不利的社区更容易出现混乱。

许多研究表明，社区失调可以预示着居民的心理健康状况较差，包括焦虑、愤怒和抑郁的发生率上升(Diez Roux，Mair，2010；Ross，2011)。例如，纽约的一项研究发现，与生活在更有序的社区受访者相比，生活在建筑质量较差的无序社区中的居民，在过去6个月内患抑郁症的概率增加29%~58%，而在他们一生中的任何时候，患抑郁症的概率增加36%~64%(Galea et al.，2005)。这些结果针对个人收入、社区收入、性别、年龄和种族进行了调整，表明社区失调的影响与社区劣势或其他社会因素是分开的。社区失调对心理健康的负面影响可能源于主观疏离感——一种与他人的隔绝感(Ross，Mirowsky，2009)。反过来，这可能源于对无力感、不信任感、社会孤立感的感知，以及社会可接受的行为准则的瓦解(也称"无规范")(Ross，2011)。因此，尽管附近有很多其他人，居民却不太可能形成创造社区感和归属感的社会联系和支持网络。

社区失调也可能预示着身体健康较差,因为缺乏安全感或危险意识可能会阻碍年轻人在户外玩耍(Molnar et al.,2004)和成年人锻炼(Fish et al.,2010)。在洛杉矶市的 2000 多名成年人的样本中,在性别、种族和慢性病的同等控制条件下,那些认为自己的社区不安全的人的体重指数,比那些认为自己的社区安全的人的体重指数(BMI)高 2.81kg/m² (95%,CI = 0.11% ~ 5.52%) (Fish et al.,2010)。同样,巴西贝洛哈里桑塔的一项研究发现,超重(BMI≥25kg/m²)的概率随着人口密度的增加而降低(反映了高人口密度与社区更大的可步行性之间的关系),但随着普查区内的社会脆弱性和凶杀率的增加而增加(校正了性别、年龄和生活方式因素)(Mendes et al.,2013)。在贝洛哈里桑塔,像在洛杉矶一样,对社区的危险感可能会阻碍步行等户外体育活动;在得克萨斯的圣安东尼奥市和英格兰西北部也报道了类似的现象(Gomez et al.,2004;Harrison et al.,2007)。有趣的是,在贝洛哈里桑塔的 3400 名成年人中,最超重的人群是 18~24 岁的年轻人,其中 77.5%的人超重(Mendes et al.,2013)。

6.3 污染和废物

人类在城市环境中的活动会产生多种类型的污染和废物,每种都会对城市居民的健康和福祉产生一系列潜在的影响。在这里,我们主要关注空气污染和固体废物(垃圾;专栏 6.3)。

💬 专栏6.3

垃圾山

城市人口产生大量垃圾,包括生活垃圾(如食物残渣、纸张、塑料和其他包装)、废金属、电子废物和危险废物。据估计,2010 年全球固体废物产量超过 350 万吨/天,预计到 2025 年将增加到 600 万吨/天,到 2100 年将增加到 1200 万吨/天(Hoornweg, Bhada-Tata, 2012;Hoornweg et al., 2013)。在一定的富裕水平下,城市居民产生的废物是乡村居民的两倍(Hoornweg et al.，2013)。人口稠密的城市集中倾倒垃圾会导致大量垃圾填埋场或垃圾山的堆积(图6.7)。著名的例子包括马尼拉附近的博雅塔斯、里约热内卢的格拉玛舒垃圾场、墨西哥城的波尼安特。垃圾山为一部分拾荒者提供了生存的机会，这些拾荒者中的成人和儿童通过垃圾分类并回收贵重物品来赚钱。

图6.7　拾荒者在巴西一座垃圾山上工作，将垃圾分类收集、回收(摄影：Marcello Casal Jr.)

拾荒者通过分拣和回收各种塑料、金属、计算机零件、打印机墨盒和其他电子废物来提供重要的服务，并且他们的收入要比包括农场、工厂在内的其他多种职业的收入高得多(Medina，2005，2008；Power，2006；Samson，2008)。现代的拾荒者的功能类似于19世纪伦敦的拾荒者，据估计19世纪50年代拾荒者的数量超过10万人(Johnson，2006)。在一些情况下，拾荒者的棚户区在垃圾山周围冒出来；这些城镇本身就藏有各种污染物和疾病，包括破伤风和肺结核。2000年，几周大雨过后，马尼拉附近的博雅塔斯垃圾场的一部分坍塌了，大量的泥石流造成200多人丧生(Power，2006)。但是，在灾难发生后的几个月内,垃圾场重新开放,并继续为成千上万的拾荒者及其家人提供生计和住所。

6.3.1 室外空气污染

城市的空气污染包含多种多样的气体混合物(如一氧化二氮、二氧化硫、一氧化碳、臭氧)以及颗粒物,这些混合气体大部分是由用于运输、工业、家庭取暖和烹饪而燃烧化石燃料产生的(参见第5章)。颗粒物本身是化合物的异质混合物,其化学组成、大小和表面积均不同(Brook, 2008)。颗粒物一般依据空气动力学直径(μm)进行分类,最小的颗粒分为细颗粒物($<2.5\mu m$;称为$PM_{2.5}$)和超细颗粒物($<0.1\mu m$;$PM_{0.1}$)。粗颗粒物的直径范围为$2.5\sim10\mu m$(PM_{10})。PM_{10}包括灰尘、花粉、真菌孢子、土壤和金属,平均寿命为数小时至数天,并且分布在离源头$10\sim100km$的地方(Brook, 2008)。$PM_{2.5}$由元素碳、碳氢化合物、有机化合物和金属组成,平均寿命为数天至数周,并且按区域(距排放源$\geqslant1000km$)分布。$PM_{0.1}$也有类似的混合化学成分,但平均寿命仅为数分钟至数小时,并且分布在离源头数百米的地方(Brook, 2008)。因此,城市中颗粒物污染的分布和构成可能有很大差异,在交通繁忙的道路附近,$PM_{0.1}$和$PM_{2.5}$的浓度最高。

短期和长期暴露于城市环境中的颗粒物都会与人患心血管疾病的风险增加有关,包括心脏缺血、心脏病发作、急性心力衰竭、中风和外周动脉疾病(Brook, 2008)。例如:暴露于交通环境中仅一小时后,心脏病发作的风险可能会增加2.7倍(Peters et al., 2004)。在一项针对美国超过6.5万名健康、绝经后的妇女的研究中,长期$PM_{2.5}$暴露量增加$10\mu g/m^3$,与心血管疾病致死风险增加76%相关(Miller et al., 2007)。暴露于颗粒物污染中,也增加了呼吸系统疾病的发病率,包括哮喘、支气管炎和肺癌,即使是在非吸烟群体中,也是如此(Valavanidis et al., 2008)。现有证据表明,细颗粒和超细颗粒($PM_{2.5}$和$PM_{0.1}$)在心血管疾病和其他疾病中的作用更大,因为这些颗粒足够小,可以深深地渗透到呼吸道中,并可能滞留在肺实质(肺中参与气体交换的部分)中(Valavanidis et al., 2008)。细颗粒物空气污染对健康的不良影响取决于大气中颗粒物的浓度和暴露时间,长期暴露会产生更大和更持久的影响(Pope, 2007)。2010年,全球约有320万(大概范围为280万~360万)人死亡和7600万人因伤残调整生命年,都归因于周围环境颗粒物污染,该风险因素引起疾病的责任在全球排名第九(Lim et al., 2012)。伤残调整生命年,指因疾病而残障的生命年数加上因过早死亡而丧失的生命年数之和(Jordan et al., 2014)。

6.3.2 室内空气污染

尽管人们对城市室外空气污染的关注更加明显,但是室内空气污染实际上是导致全球疾病增加的更大因素。2010年,化石燃料燃烧造成的家庭空气污染导致全球350万人死亡和1.08亿伤残调整生命年;该危险因素在全球致病因素中排名第三,仅次于高血压和吸烟(包括二手烟)(Lim et al., 2012)。就区域而言,家庭空气污染在亚洲和撒哈拉以南非洲造成了最大的健康代价,在这些地区中,木材、木炭、粪便和农作物残渣等生物质燃料通常被用于烹饪,许多住宅的通风不良(Saksena et al., 2003; Bailis et al., 2005; Dasgupta et al., 2006; Zhou et al., 2011)。如果不切实改变做法(如向木炭或液化石油气等燃烧更清洁的燃料过渡),预计2000年至2030年之间,使用生物质燃料导致的室内空气污染,可导致非洲980万人过早死亡(Bailis et al., 2005)。该预测值包括因下呼吸道感染导致的810万5岁以下儿童死亡和170万成年女性死于慢性阻塞性肺疾病。

在发达国家中,甲醛和挥发性有机化合物(VOCs)等家用和办公用品有害排放物引起的室内污染越来越受到人们的关注(Zhang, Smith, 2003; Steinemann, 2009)。挥发性有机化合物是具有低沸点的有机化学物质,在室温下具有较高的蒸气压,因此大量分子会蒸发(从液体源)或升华(从固体源)进入空气。这些化合物通常存在于油漆、胶水、塑料、合成纤维和溶剂中,并且室内的浓度可能高于室外(Mølhave, 1986; Dales et al., 2008)。已知挥发性有机化合物(如苯和1,3-丁二烯)对人体有致癌作用,室内接触挥发性有机化合物会增加患白血病和淋巴瘤的相对风险(Viegi et al., 2004; Irigarary et al., 2007)。芳香家居产品,如空气清新剂、洗洁精、洗衣粉以及织物柔软剂,也包含多种VOCs,其中一些被归类为有毒或有害物质,如乙醛、氯甲烷(Steinemann, 2009)。使用芳香产品会导致各种健康问题,包括头痛、哮喘和接触过敏,但其中所含和释放的大多数挥发性有机化合物并未列在产品成分中,因此难以识别有问题的成分(Steinemann et al.，2011）。

6.4 城市环境中的气候变化

城市热岛效应会加剧夏季热浪期间的白天和夜间的高温，使城市居民的生活环境更加恶劣和危险。坚硬的城市表面在白天会迅速升温，然后在夜间长时间释放热量，这使城市居民在炎热的天气中承受持续的热压力，并增加发病率和死亡率（Weng，Yang，2003；Harlan et al.，2006；Kalkstein et al.，2013；Azhar et al.，2014；Hondula，Barnett，2014）。在热浪期间，许多不同原因的死亡频率增加，包括心脏骤停、中风和呼吸衰竭（Kovats，Hajat，2008；Giua et al.，2010；Tan et al.，2010；Chen et al.，2013；Kalkstein et al.，2013）。

例如，在1998年8月的9天热浪中，中国上海的城市热岛效应导致了大量与高温相关的死亡（Tan et al.，2010）。上海市中心的全因死亡率超过27.3/10万人，而远郊区则是7/10万人。随着热浪的持续，每天的额外死亡人数增加，在第九天也是最后一天达到453人的峰值（Tan et al.，2010）。在澳大利亚布里斯班进行的一项研究发现，夏季每天最高气温升高10℃之后，当天医院人数增加了7.2%（95%，CI=4.7%~9.8%）（Hondula，Barnett，2014）。而另一位观察者发现，每天最高气温每升高1℃，出血性中风的住院率就会增加15%（5%~26%）（Wang et al.，2009）。由于城市温度升高，某些城市人群中包括老年人、社会弱势群体和无家可归者，患病和死亡的风险最大（Harlan et al.，2006；Kovats，Ebi，2006；Kovats，Hajat，2008；Loughnan et al.，2012）。

高温还会影响城市中的人类行为；现场研究和实验室试验表明，高温会增加易怒性和攻击性（有时称为热效应）（Anderson，2001）。暴力犯罪随着温度的升高而增加（Anderson，2001；Anderson，Delisi，2011）。根据1950—2008年间美国50个城市的数据，在较热的年份，暴力犯罪（谋杀和严重袭击）的发生率较高，每年平均增加1℃，平均每10万人就增加142起犯罪（相当于每上升1°F，即每10万人中有79宗犯罪）（Anderson，DeLisi，2011）。犯罪也随着气温的不同而不同。在密苏里州圣路易斯，每月气温比正常气温升高1℃，每月平均增加9.9起暴力犯罪（或1.3%）（Mares，2013）。圣路易斯在最冷和最热的月份(1月和8月)之间的正常温度变化接近28℃，也就是说，预计暴力犯罪的季节性波动约为35%。

在气候变化的影响下，极端热浪及其对城市人口的不利影响预计会增加，包括热浪的数量和持续时间(IPCC，2014)，以及攻击性和犯罪率(Anderson，DeLisi，2011；Mares，2013)。叠加城市的热岛效应，更频繁的热浪将导致世界上许多城市中与高温相关的疾病和死亡率增加(Huang et al.，2011a；IPCC，2014)。此外，在美国年平均气温升高2℃(3.6℉)(这是预计2025年北美大部分地区的可能状况；IPCC，2014)，每年可能导致增加5万起暴力犯罪(谋杀和严重袭击)(Anderson，DeLisi，2011)。除了巨大的个人和社会成本外，这种疾病、暴力和死亡额外负担的经济成本也很高。据估计，暴力犯罪给社会造成的损失为每起谋杀案128万美元，每起严重袭击案19537美元(McCollister et al.，2010)。因此，如果在美国每年因平均气温升高2℃而造成的额外暴力犯罪中，有10%是谋杀案，这将给社会带来72.8亿美元的额外成本。

6.5 世界城市的卫生不平等

尽管生活在城市地区的人往往比乡村地区的人拥有更好的健康状况和寿命，但是这些广泛的趋势，掩盖了整个城市范围内居民卫生的严重不平等现象。例如，城市贫困妇女在分娩期间缺乏熟练的接生员，从而发生可能导致产妇残疾或死亡的并发症的风险更高(WHO，UN-Habitat，2010)。在孟加拉国，具有熟练技术的接生员的服务范围，从6%的最贫穷城市妇女到75%的最富有妇女。一项针对21个国家/地区的研究表明，城市女性感染艾滋病毒的可能性比城市男性高50%，比乡村女性高80%，其社会经济地位低下加剧了这一问题(WHO，UN-Habitat，2010)。根据《人口与健康调查》(2003—2008年)的最新数据，非洲斯威士兰超过35%的城市妇女是HIV阳性(WHO，UN-Habitat，2010)。

家庭财富在世界各地城市的儿童营养不良以及婴幼儿死亡率方面也起着重要作用。在全球范围内，慢性营养不良导致儿童期死亡的比例占1/3以上；尽管在城市中5岁以下儿童的营养不良现象比乡村地区少，但在整个城市人口中却有很大差异。在非洲、亚洲和美洲，最贫穷的城市儿童因营养不良而发育迟缓的概率是最富有的儿童的三倍;随着家庭收入的增加,儿童慢性营养不良的风险逐渐降低(WHO, UN-Habitat, 2010)。在这三个地区中,城市最贫穷家庭的儿童在5岁前死亡的概率约是城市最富裕家庭儿童的两倍。非洲的儿童死亡率最高,城市最贫困家庭的孩子活产率大于130/1000(数据范围为2000—2007年)。

城市贫民的卫生负担——由妇女和儿童不成比例地承担——有一系列原因,包括住房质量差、卫生条件不足和无法获得清洁饮用水;贫民窟和其他非正规居住区的人口密度很高,促进了人与人之间接触传播的传染病,如下呼吸道感染和结核病;难以获得医疗保健服务,较高的暴力犯罪率和较高的冒险行为发生率(Zulu et al., 2002; Campbell, Campbell, 2007; COHRE, 2008; Montgomery, 2009; WHO, UN-Habitat, 2010; Mendes et al., 2013)。非正规居住区的住宅和共用厕所设施不安全,使妇女遭受性侵犯和感染包括艾滋病毒在内的性传播疾病的风险增加(Kalichman, Simbayi, 2004; Amuyunzu-Nyamongo et al., 2007; COHRE, 2008; Amnesty International, 2010)。贫穷、无家可归和食物不足,也可能导致妇女以性行为换食物或住宿,这反过来又增加了她们感染性传播疾病的风险(Weiser et al., 2007)。

贫民窟和其他非正规居住区的居民的住房使用权极不确定,而且有可能因"清理贫民窟"而被逐出家园(Greene, 2003; Padhi, 2007; COHRE, 2008; Amnesty International, 2010; Shin, Li, 2012; Steinbrink, 2013)。最近一个值得注意的居民从非正规居住区被强制驱逐的例子是津巴布韦总统穆加贝(Mugabe)于2005年发起的"净化运动"(Operation Murambatsvina,被称为恢复秩序行动或清除垃圾行动)。该行动由警察和军队迅速清除贫民窟,首先在哈拉雷实施,然后是全国其他城市。估计有70万人在行动期间失去了家园、生计来源或者两者兼而有之,这使数十万城市贫困人口陷入了更加贫困的境地(Tibaijuka, 2005)。清除贫民窟常常与大型世界赛事活动(如奥运会和足球世界杯)的建设项目相关,并渴望"清理"(塑造)该赛事主办城市的形象(Greene, 2003; Power, 2006; Shin, Li, 2012)。城市贫民也往往最易遭受飓风和洪灾等自然灾害的影响,随着气候变化的加剧,这一趋势也将会加剧(Anderson, DeLisi, 2011; IPCC, 2014)。

6.6 总结

城镇化进程对城市居民产生了广泛的影响，从个体到家庭、社区和整个人口。尽管城市可以为居民提供很多，包括机会、就业、文化、高效能的生活和社区意识，但是城市也可以成为肮脏、犯罪、疾病和绝望的地方。城市形态的特征可以促进或阻碍人类的健康和福祉。社区设计和秩序感或无秩序感，会影响人的运动方式和超重、肥胖、社会联系、社交隔离和抑郁程度。进入社区公园锻炼身体，可以改善身心健康，同时促进城市居民与自然之间的持续联系；城市扩张增加了人民对汽车的依赖，并且可以造成与社区隔离的感觉。城市产生大量的污染物、废物和热量，也可能促进传染病的持续和传播。在世界各地的城市中，正是城市贫困与这些令人不愉快的城镇化产物生活得最密切；在城市社会中，优势群体和弱势群体的健康状况之间形成鲜明对比。但是，城市是现在和未来的发展方向，不可能回归农业生活方式。我们必须调整我们的城市，以应对污染、浪费、高温、无序扩张，以及与自然隔绝的问题，而不仅仅是富裕社区的居民和连通性很好的社区居民。在下一章中将讨论为人类以及与我们共享城市的所有其他物种，创建更健康、更宜居的城市空间的实用策略。

思考题

1. 人类在城市中体验自然的机会很少。你如何在你熟悉的城市或乡镇中增加这些机会？
2. 讨论进入城市中的公园和其他绿色开敞空间，如何对人类健康至关重要。
3. 举例说明，在城市环境中发生人类与野生动物相互作用的范围。
4. 定义以下术语：步行性、城市扩张和社区失调。它们是如何影响城市人类的健康和福祉的？
5. 与城镇化相关的气候变化，是如何影响城市居民的？
6. 在城市中室内和室外空气污染对人类健康的最大威胁是什么？为什么？
7. 描述造成世界城市卫生不平等的主要因素。谁承担这种不平等的最大责任？

本章参考文献

Adams A L, Dickinson K J M, Robertson B C, et al. (2013) Predicting summer site occupancy for an invasive species, the common brushtail possum (*Trichosurus vulpecula*), in an urban environment. *PLoS ONE* 8, e58422.

Allen A P, O'Conner, R. J. (2000) Hierarchical correlates of bird assemblage structure on northeastern U.S.A. lakes. *Environmental Monitoring and Assessment*, 62, 15-37.

Amnesty International. (2001) *Insecurity and Indignity: Women's Experiences in the Slums of Nairobi, Kenya*. Amnesty International Publications, London.

Amuyunzu-Nyamongo M, Okeng'o L, Wagura A, et al. (2007) Putting on a brave face: the experience of women living with HIV and AIDS in informal settlements of Nairobi, Kenya. *AIDS Care*, 19, 25-34.

Anderson C A. (2001) Heat and violence. *Current Directions in Psychological Science*, 10, 33-38.

Anderson C A, DeLisi M. (2011) Implications of global climate change for violence in developed and developing countries. In *The Psychology of Social Conflict and Aggression*. Forgas J, Kruglanski A, Williams K eds. Psychology Press, New York, 249-265.

Azhar G S, Mavalankar D, Nori-Sarma A, et al. (2014) Heat-Related mortality in India: excess all-cause mortality associated with the 2010 Ahmedabad Heat Wave. *PLoS ONE*, 9(e), 91831.

Bailis R, Ezzati M, Kammen D M. (2005) Mortality and greenhouse gas impacts of biomass and petroleum energy futures in Africa. *Science*, 308, 98-103.

Banks W A, Altmann J, Sapolsky R M, et al. (2003) Serum leptin levels as a marker for a syndrome X-like condition in wild baboons. *The Journal of Clinical Endocrinology and Metabolism*, 88, 1234-1240.

Barton J, Pretty J. (2010) What is the best dose of nature and green exercise for improving mental health? A multi-study analysis. *Environmental Science and Technology*, 44, 3947-3955.

Bateman P W, Fleming P A. (2012) Big city life: carnivores in urban environments. *Journal of Zoology*, 287, 1-23.

Bates D G. (2012) On forty years: remarks from the editor. *Human Ecology*, 40, 1-4.

Bauman A E, Reis R S, Sallis J F, et al. (2012) Correlates of physical activity: why are some people physically active and others not? *The Lancet*, 380, 258-271.

Beckmann J P, Berger J. (2003) Rapid ecological and behavioural changes in carnivores: the response of black bears (*Ursus americanus*) to altered food. *The Zoological Society of London*, 261, 207-212.

Beier P. (1991) Cougar attacks on humans in the United States and Canada. *Wildlife Society Bulletin*, 19, 403-412.

Bhatia S, Athreya V, Grenyer R, et al. (2013) Understanding the role of representations of human-leopard conflict in Mumbai through media-content analysis. *Conservation Biology*, 27, 588-594.

Brook R D. (2008) Cardiovascular effects of air pollution. *Clinical Science*, 115, 175-187.

Campbell M. (2008) An animal geography of avian feeding habits in Peterborough, Ontario. *Area*, 40, 472-480.

Campbell T, Campbell A. (2007) Emerging disease burdens and the poor in cites of the developing world. *Journal of Urban Health: Bulletin of the New York Academy of Medicine*, 84, 54-64.

Carrus G, Scopelliti M, Lafortezza R, et al. (2015) Go green, feel better? The positive effects of biodiversity on the well-being of individuals visiting urban and peri-urban green areas. *Landscape and Urban Planning*, 134, 221-228.

Casagrande S S, Gittelsohn J, Zonderman A B, et al. (2011) Association of walkability with obesity in Baltimore City, Maryland. *American Journal of Public Health*, 101, 318-324.

Chamberlain D E, Vickery J A, Glue D E, et al. (2005) Annual and seasonal trends in the use of garden feeders by birds in winter. *Ibis*, 147, 563-575.

Chapman R, Jones D N. (2009) Just feeding the ducks: quantifying a common wildlife-human interaction. *The Sunbird*, 39, 19-28.

Chapman R, Jones D N. (2011) Foraging by native and domestic ducks in urban lakes: behavioural implications of all that bread. *Corella*, 35, 101-106.

Cervero R, Kockelman K. (1997) Travel demand and the 3Ds: density, diversity, design. *Transportation Research Part D—Transport and Environment*, 2, 199-219.

Cervero R, Sarmiento O L, Jacoby E, et al. (2009) Influences of built environments on walking and cycling: lessons from Bogota. *International Journal of Sustainable Transportation*, 3, 203-226.

Chaudhry P, Bagra K, Singh B. (2011) Urban greenery status of some Indian cities: a short communication. *International Journal of Environmental Science and Development*, 2, 98-101.

Chen R, Wang C, Meng X, et al. (2013) Both low and high temperature may increase the risk of stroke mortality. *Neurology*, 81, 1064-1070.

Chiesura A. (2004) The role of urban parks for the sustainable city. *Landscape and Urban Planning*, 68, 129-138.

Choudhury A. (2004) Human-elephant conflicts in Northeast India. *Human Dimensions of Wildlife*, 9, 261-270.

Cilliers J. (2008) *Africa in the New World: How Global and Domestic Developments will Impact by* 2025. Institute for Security Studies Monographs, no. 151. Institute for Security Studies, Pretoria.

Clark N E, Lovell R, Wheeler B W, et al. (2014) Biodiversity, culture pathways, human health: a framework. *Trends in Ecology and Evolution*, 29, 198-204.

Clarke L W, Jenerette G D, Davalia A. (2013) The luxury of vegetation and the legacy of tree biodiversity in Los Angeles, CA. *Landscape and Urban Planning*, 116, 48-59.

COHRE. (2008) *Women, Slums and Urbanization: Examining the Causes and Consequences*. Centre on Housing Rights and Evictions, Geneva.

Coley R L, Kuo F E, Sullivan W C. (1997) Where does community grow? The social context created by nature in urban public housing. *Environment and Behavior*, 29,

468-494.

Dales R, Liu L, Wheeler A J, et al. (2008) Quality of indoor residential air and health. *Canadian Medical Association Journal*, 179, 147-152.

Daniels G D, Kirkpatrick J B. (2006) Does variation in garden characteristics influence the conservation of birds in suburbia? *Biological Conservation*, 133, 326-335.

Dasgupta S, Huq M, Khaliquzzaman M, et al. (2006) Indoor air quality for poor families: new evidence from Bangladesh. *Indoor Air*, 16, 426-444.

Dieleman F, Wegener M. (2004) Compact city and urban sprawl. *Built Environment*, 30, 308-323.

Diez Roux A V, Mair C. (2010) Neighborhoods and health. *Annals of the New York Academy of Sciences*, 1186, 125-145.

Escobedo F J, Nowak D J, Wagner J E, et al. (2006) The socioeconomics and management of Santiago de Chile's public urban forests. *Urban Forestry and Urban Greening*, 4, 105-114.

Ewing R, Pendall R, Chen D. (2003) Measuring sprawl and its transportation impacts. *Transportation Research Record: Journal of the Transportation Research Board*, 1, 175-183.

Farr D. (2012) *Sustainable Urbanism: Urban Design with Nature*. John Wiley & Sons Inc, Hoboken.

Fish J S, Ettner S, Ang A, et al. (2010) Association of perceived neighbourhood safety on body mass index. *American Journal of Public Health*, 100, 2296-2303.

Frank L D, Anderson M A, Schmid T L. (2004) Obesity relationships with community design, physical activity, time spent in cars. *American Journal of Preventive Medicine*, 27, 87-96.

Fuller R A, Irvine K N, Devine-Wright P, et al. (2007) Psychological benefits of greenspace increase with biodiversity. *Biology Letters*, 3, 390-394.

Fuller R A, Warren P H, Armsworth P R, et al. (2008) Garden bird feeding predicts the structure of urban avian assemblages. *Diversity and Distributions*, 14, 131-137.

Galea S, Ahern J, Rudenstine S, et al. (2005) Urban built environment and depression: a multilevel analysis. *Journal of Epidemiology and Community Health*, 59, 822-827.

Garden F L, Jalaludin B B. (2009) Impact of urban sprawl on overweight, obesity, physical activity in Sydney, Australia. *Journal of Urban Health*, 86, 19-30.

Germano J M, Bishop P J. (2009) Suitability of amphibians and reptiles for translocation. *Conservation Biology*, 23, 7-15.

Giles-Corti B, Broomhall M H, Knuiman M, et al. (2005) Increasing walking. How important is distance to, attractiveness, size of public open space? *American Journal of Preventive Medicine*, 28, 169-176.

Giles-Corti B, Macaulay G, Middleton N, et al. (2015) Developing a research and practice tool to measure walkability: a demonstration project. *Health Promotion Journal of Australia*, 25, 160-166.

Giua A, Abbas M A, Murgia N, et al. (2010) Climate and stroke: a controversial association. *International Journal of Biometeorology*, 54, 1-3.

Glue D. (2006) Variety at winter bird tables. *Bird Populations*, 7, 212-215.

Gomez J E, Johnson B A, Selva M, et al. (2004) Violent crime and outdoor activity among inner-city youth. *Preventive Medicine*, 39, 876-881.

Grahn P, Stigsdotter U A. (2003) Landscape planning and stress. *Urban Forestry and Urban Greening*, 2, 1-18.

Green M. (2014) Little Fox, Big Problem. *The Age*, July 4th, 2014. http://www.theage.com.au/action/printArticle? id=5567201.

Greene S J. (2003) Staged cities: mega-events, slums clearance, global capital. *Yale Human Rights and Development Law Journal*, 6, 161-187.

Hallal P C, Andersen L B, Bull F, et al. (2012) Global physical activity levels: surveillance progress, pitfalls, prospects. *The Lancet*, 380, 247-257.

Hammer M S, Swinburn T K, Neitzel R L. (2014) Environmental noise pollution in the United States: developing an effective public health response. *Environmental Health Perspectives*, 122, 115-119.

Hankey S, Marshall J D, Brauer M. (2012) Health impacts of the built environment:

within-urban variability in physical inactivity, air pollution, ischemic heart disease mortality. *Environmental Health Perspectives*, 120, 247-253.

Haq S M A. (2011) Urban green space and an integrative approach to sustainable environment. *Journal of Environmental Protection*, 2, 601-608.

Harlan S L, Brazel A J, Prashad L, et al. (2006) Neighborhood microclimates and vulnerability to heat stress. *Social Science and Medicine*, 63, 2847-2863.

Harper M J, McCarthy M A, van der Ree R. (2008) Resources at the landscape scale influence possum abundance. *Austral Ecology*, 33, 243-252.

Harrison R A, Gemell I, Heller R F. (2007) The population effect of crime and neighbourhood on physical activity: an analysis of 15,461 adults. *Journal of Epidemiology and Community Health*, 61, 34-39.

Hes D, Plessis C D. (2015) *Designing for Hope: Pathways to Regenerative Sustainability*. Rout-ledge, Abingdon.

Heynen N C. (2003) The scalar production of injustice within the urban forest. *Antipode*, 35, 980-998.

Hondula D M, Barnett A G. (2014) Heat-related morbidity in Brisbane, Australia: spatial variation and area-level predictors. *Environmental Health Perspectives*, 122, 831-836.

Hoornweg D, Bhada-Tata P. (2012) *What a Waste: A Global Review of Solid Waste Management*. World Bank, Washington DC. http://siteresources.worldbank.org/INTURBANDEVELOPMENT/Resources/336387-1334852610766/What_a_Waste2012_Final.pdf(accessed 10/06/2015).

Hoornweg D, Bhada-Tata P, Kennedy C. (2013) Waste production must peak this century. *Nature*, 502, 615-617.

Huang C, Barnett A G, Wang X, et al. (2011a) Projecting future heat-related mortality under climate change scenarios: a systematic review. *Environmental Health Perspectives*, 119, 1681-1690.

Huang G, Zhou W, Cadenasso M L. (2011b) Is everyone hot in the city? Spatial pattern of land surface temperatures, land cover, neighborhood socioeconomic characteristics in Baltimore, MD. *Journal of Environmental Management*, 92, 1753-1759.

IPCC. 2014. *Climate Change* 2014: *Synthesis Report. Contribution of Working Groups I, II and III to the Fifth Assessment Report of the Intergovernmental Panel on Climate Change.* Pachauri R K, Meyer L A eds. IPCC, Geneva.

Irigarary P, Newby J A, Clapp R, et al. (2007) Lifestyle-related factors and environmental agents causing cancer: an overview. *Biomedicine and Pharmacotherapy* 61, 640-658.

Johnson S. (2006) *The Ghost Map: The Story of London's Most Terrifying Epidemic, How it Changed Science, Cities, the Modern World.* Riverhead, New York.

Jones D N, Reynolds S J. (2008) Feeding birds in our towns and cities: a global research opportunity. *Journal of Avian Biology*, 39, 265-271.

Jordan H, Dunt D, Hollingsworth B, et al. (2014) Costing the morbidity and mortality consequences of zoonoses using health-adjusted life years. *Transboundary and Emerging Diseases.* doi: 10.1111/tbed.12305.

Kaczynski A T, Henderson K A. (2007) Environmental correlates of physical activity: a review of evidence about parks and recreation. *Leisure Science*, 29, 315-354.

Kalichman S C, Simbayi L C. (2004) Sexual assault history and risk for sexually transmitted infections among women in an African township in Cape Town, South Africa. *AIDS Care*, 16, 681-689.

Kalkstein L S, Sailor D, Shickman K, et al. (2013) *Assessing the health impacts of urban heat island reduction strategies in the District of Columbia.* Global Cool Cities Alliance, Washington DC. http://www.coolrooftoolkit.org/wp-content/uploads/2013/12/DC-Heat-Mortality-Study-for-DDOE-FINAL.pdf(accessed on 12/06/2015).

Kennedy C, Cuddihy J, Engel-Yan J. (2007) The changing metabolism of cities. *Journal of Industrial Ecology*, 11, 43-59.

Kennedy C, Pincetl S, Bunje P. (2011) The study of urban metabolism and its applications to urban planning and design. *Environmental Pollution*, 159, 1965-1973.

Kim D. (2008) Blues from the neighborhood? Neighborhood characteristics and depression. *Epidemiologic Reviews*, 30, 101-107.

Kingsley J Y, Townsend M. (2006) "Dig in" to social capital: community gardens as mech-anisms for growing urban social connectedness. *Urban Policy and Research*, 24, 525-537.

Knopfler M. (1982) *Telegraph Road*. Vertigo/Universal Music UK, London.

Kovats R S, Ebi K L. (2006) Heatwaves and public health in Europe. *European Journal of Public Health*, 16, 592-599.

Kovats R S, Hajat S. (2008) Heat stress and public health: a critical review. *The Annual Review of Public Health*, 29, 1-9.

Kuo F E. (2001) Coping with poverty: impacts of environment and attention in the inner city. *Environment and Behavior*, 33, 5-34.

Lamarque F, Anderson J, Fergusson R, et al. (2008) *Human-Wildlife Conflict in Africa: Causes, Consequences, Management Strategies*. FAO Forestry Paper 157. Food and Agriculture Orga-nization of the United Nations, Rome. http://www.fao.org/docrep/012/i1048e/i1048e00.htm(accessed 10/06/2015).

Lee K E, Williams K J H, Sargent L D, et al. (2015) 40-second green roof views sustain attention: the role of micro-breaks in attention restoration. *Journal of Environmental Psychology*, 42, 182-189.

Lim S S, Vos T, Flaxman A D, et al. (2012) A comparative risk assessment of burden of disease and injury attributable to 67 risk factors and risk factor clusters in 21 regions, 1990-2010: a systematic analysis for the Global Burden of Disease Study 2010. *Lancet*, 380, 2224-2260.

Linnell J D, Aanes R, Swenson J E. (1997) Translocation of carnivores as a method for managing problem animals: a review. *Biodiversity and Conservation*, 6, 1245-1257.

Lopez R. (2004) Urban sprawl and risk for being overweight or obese. *American Journal of Public Health*, 94, 1574-1579.

Loughnan M, Nicholls N, Tapper N J. (2012) Mapping heat health risks in urban areas. *International Journal of Population Research*, 2010, 1-12.

Lowe M, Boulange C, Giles-Corti B. (2014) Urban design and health: progress to date and future challenges. *Health Promotion Journal of Australia*, 25, 14-18.

Maas J, van Dillen S M E, Verheij R A, et al. (2009) Social contact as a possible mechanism behind the relationship between green space and health. *Health and Place*, 15, 586-595.

Maller C, Townsend M, Pryor A, et al. (2004) Healthy nature healthy people: "Contact with nature" as an upstream health promotion intervention for populations. *Health Promotions International*, 21, 45-54.

Mares D. (2013) Climate change and crime: monthly temperature and precipitation anomalies and crime rates in St. Louis, MO 1990-2009. *Crime, Law, Social Change*, 59, 185-208.

Martinez-Fernandez C, Audirac I, Fol S, et al. (2012) Shrinking cities: urban challenges of globalization. *International Journal of Urban and Regional Research*, 36, 213-225.

Mason P R. (2009) Zimbabwe experiences the worst epidemic of cholera in Africa. *The Journal of Infection in Developing Countries*, 3, 148-151.

McCollister K E, French M T, Fang H. (2010) The cost of crime to society: new crime-specific estimates for policy and program evaluation. *Drug and Alcohol Dependence*, 108, 98-109.

McKee D. (2003) Cougar attacks on humans: a case report. *Wilderness and Environmental Medicine*, 14, 169-173.

Medina M. (2005) Serving the unserved: Informal refuse collection in Mexican cities. *Waste Management and Research*, 23, 390-397.

Medina M. (2008) The informal recycling sector in developing countries: organizing waste pickers to enhance their impact. *Gridlines Note* no. 44. Public-Private Infrastructure Advisory Facility(PPIAF), World Bank, Washington DC. https://www.ppiaf.org/sites/ppiaf.org/files/publication/Gridlines-44-Informal%20Recycling%20-%20MMedina.pdf(accessed 12/06/2015).

Mendes L L, Nogueira H, Padez C, et al. (2013) Individual and environmental factors associated for overweight in urban population of Brazil. *BMC Public Health*, 13, 988. doi: 10.1186/1471-2458-13-988.

Miller J R. (2005) Biodiversity conservation and the extinction of experience. *Trends in Ecology and Evolution*, 20, 430-343.

Miller K A, Siscovick D S, Sheppard L, et al. (2007) Long-term exposure to air pollution and incidence of cardiovascular events in women. *The New England Journal of Medicine*, 356, 447-458.

Mølhave L. (1986) Indoor air quality in relation to sensory irritation due to volatile organic compounds. *ASHRAE Transactions*, 92, 306-316.

Molnar B E, Gortmaker S L, Bull F C, et al. (2004) Unsafe to play? Neighborhood disorder and lack of safety predicts reduced physical activity among urban children and adolescents. *American Journal of Health Promotion*, 18, 378-379.

Montgomery M. (2009) Urban poverty and health in developing countries. *Population Bulletin*, 64(2).

Mukandavire Z, Liao S, Wang J, et al. (2011) Estimating the reproductive numbers for the 2008-2009 cholera outbreaks in Zimbabwe. *Proceedings of the National Academy of Sciences of the United States of America*, 108, 8767-8772.

NSW Government. (2004) *Planning Guidelines for Walking and Cycling*. Department of Infrastructure, Planning, Natural Resources, Sydney. http://www.planning.nsw.gov.au/plansforaction/pdf/guide_pages.pdf(accessed 15/06/2015).

Padhi R. (2007) Forced evictions and factory closures: rethinking citizenship and rights of working class women in Delhi. *Indian Journal of Gender Studies*, 14, 74-92.

Parsons H, Major R E, French K. (2006) Species interactions and habitat associations of birds inhabiting urban areas of Sydney, Australia. *Austral Ecology*, 31, 217-227.

Pedlowski M A, Da Silva V A C, Adell J J C, et al. (2002) Urban forest and environmental inequality in Campos dos Goytacazes, Rio de Janeiro, Brazil. *Urban Ecosystems*, 6, 9-20.

Peters A, von Klot S, Heier M, et al. (2004) Exposure to traffic and the onset of myocardial infarction. *The New England Journal of Medicine*, 351, 1721-1730.

Pietsch R S. (1994) The fate of urban Common Brushtail Possums translocated to sclerophyll forest. In *Reintroduction Biology of Australian and New Zealand Fauna* (ed M. Serena), Surrey Beatty and Sons, Chipping Norton, 239-246.

Pikora T, Giles-Corti B, Bull F, et al. (2003) Developing a framework for assessment of the environmental determinants of walking and cycling. *Social Science and Medicine*, 56, 1693-1703.

Pope A C. (2007) Mortality effects on longer term exposures to fine particulate air pollution: review of recent epidemiological evidence. *Inhalation Toxicology*, 19, 33-38.

Power M. (2006) The magic mountain: trickle-down economics in a Philippine garbage dump. *Harpers*, 1879, 57-71.

Pyle R M. (1978) The extinction of experience. *Horticulture*, 56, 64-67.

Redman C L, Jones N S. (2005) The environmental, social, health dimensions of urban expansion. *Population and Environment*, 26, 505-520.

Redman C L. (1999) *Human Impacts on Ancient Environments*. University of Arizona Press, Tucson.

Ross C E. (2011) Collective threat, trust, the sense of personal control. *Journal of Health and Social Behaviour*, 52, 287-296.

Ross C E, Mirowsky, J. (2009) Neighborhood disorder, subjective alienation, distress. *Journal of Health and Social Behavior*, 50, 49-64.

Saldivar-Tanaka L, Krasny M E. (2004) Culturing community development, neighborhood open space, civic agriculture: the case of Latino community gardens in New York City. *Agriculture and Human Values*, 21, 399-412.

Saelens B E, Sallis J F, Frank L D. (2003) Environmental correlates of walking and cycling: findings from the transportation, urban design, planning literatures. *Environment and Physical Activity*, 25, 80-91.

Saksena S, Singh P B, Kumar Prasad R, et al. (2003) Exposure of infants to outdoor and indoor air pollution in low-income urban areas-a case study of Delhi. *Journal of Exposure Analysis and Environmental Epidemiology*, 13, 219-230.

Šálek M, Drahnikova L, Tkadlec E. (2015) Changes in home-range sizes and population densities of carnivore species along the natural to urban habitat gradient. *Mammal Review*, 45, 1-14.

Sallis J, Millstein R, Carlson J. (2011) Community design for physical activity. In *Making Healthy Places: Designing and Building for Health, Well-Being, Sustainability*. Dannen-berg A, Frumkin H, Jackson R eds. Island Press, Washington DC, 33-49.

Sallis J F, Saelens B E, Frank L D, et al. (2009) Neighborhood built environment and income: examining multiple health outcomes. *Social Science & Medicine*, 68, 1285-1293.

Samson M. (2008) *Refusing to be Cast Aside: Waste Pickers Organising around the World*. Women in Informal Employment: Globalizing and Organizing (WIEGO), Cambridge, Massachusetts.

Saunders G, Coman B, Kinnear J, et al. (1995) *Managing Vertebrate Pests: Foxes*. Bureau of Resource Sciences, Australian Government, Canberra. http://www.southwestnrm.org.au/sites/default/files/uploads/ihub/saunders-g-et-al-1995-managing-vertebrate-pests-foxes-table-contents.pdf (accessed (15/04/2014).

Schell L M, Denham M. (2003) Environmental pollution in urban environments and human biology. *Annual Review of Anthropology*, 32, 111-134.

Silver C. (1997) The racial origins of zoning in American cities. In *Urban Planning and the African American Community: In the Shadows*. Thomas J M, Ritzdorf M eds. Sage Publications, Inc, Thousand Oaks, 23-42.

Shanahan D F, Fuller R A, Bush R, et al. (2015) The health benefits of urban nature: how much de we need? *BioScience*, 65, 476-485.

Shanahan D F, Lin B B, Gaston K J, et al. (2011) Socio-economic inequalities in access to nature on public and private lands: a case study from Brisbane, Australia. *Landscape and Urban Planning*, 130, 14-23.

Shin H B, Li B. (2012) Migrants, landlords and their uneven experiences of the Beijing Olympic Games, in Centre for Analysis of Social Exclusion Paper 163.

London School of Economics, London.

Smith K R, Brown B B, Yamada I, et al. (2008) Walkability and body mass index: density, design, new diversity measures. *American Journal of Preventive Medicine*, 35, 237-244.

Soule D. (2009) *Urban Sprawl: A Comprehensive Reference Guide*. Greenwood Press, West-port.

Spencer R D, Beausoleil R A, Martorello D A. (2007) How agencies respond to human-black bear conflicts: a survey of wildlife agencies in North America. *Ursus*, 18, 217-229.

Steinbrink M. (2013) Festifavelisation: mega-events, slums and strategic city-staging-the example of Rio de Janeiro. *Journal of the Geographical Society of Berlin*, 144, 129-145.

Steinemann A C. (2009) Fragranced consumer products and undisclosed ingredients. *Environmental Impact Assessment Review*, 29, 32-38.

Steinemann A C, MacGregor I C, Gordon S M, et al. (2011) Fragranced consumer products: chemicals emitted, ingredients unlisted. *Environmental Impact Assessment Review*, 31, 328-333.

Sugiyama T, Francis J, Middleton N J, et al. (2010) Associations between recreational walking and attractiveness, size, and proximity of neighborhood open spaces. *American Journal of Public Health*, 100, 1752-1757.

Sugiyama T, Neuhaus M, Cole R, et al. (2012) Destination and route attributes associated with adults' walking: a review. *Medicine and Science in Sports and Exercise*, 44, 1275-1286.

Sullivan B K, Kwiatkowski M A, Schuett G W. (2004) Translocation of urban Gila Monsters: a problematic conservation tool. *Biological Conservation*, 117, 235-242.

Tan J, Zheng Y, Tang X, et al. (2010) The urban heat island and its impact on heat waves and human health in Shanghai. *International Journal of Biometeorology*, 54, 75-84.

The Editors. (1972) Introductory statement. *Human Ecology*, 1, 1.

Thornton C, Quinn M S. (2009) Coexisting with cougars: public perceptions, attitudes, awareness of cougars on the urban-rural fringe of Calgary, Alberta, Canada. *Human-Wildlife Conflicts*, 3, 282-295.

Tibaijuka A K. (2005) *Report of the Fact-finding Mission to Zimbabwe to Assess the Scope and Impact of Operation Murambatsvina by the UN Special Envoy on Human Settlement Issues in Zimbabwe*. UN-Habitat, Nairobi. http://www1.umn.edu/humanrts/research/ZIM%20UN%20Special%20Env%20Report.pdf (accessed 12/06/2015).

Turner W R, Nakamura T, Dinetti M. (2004) Global urbanization and the separation of humans from nature. *BioScience*, 54, 585-590.

Ulrich R S. (1984) Human response to vegetation and landscapes. *Landscapes and Urban Planning*, 13, 29-44.

UN-Habitat. (2003) *Slums of the World: The Face of Urban Poverty in the New Millennium?* UN-Habitat, Nairobi. http://www.sustainable-design.ie/sustain/UN-Habitat_2003WorldSlums Report.pdf (accessed 12/06/2015).

Valavanidis A, Fiotakis K, Vlachogianni T. (2008) Airborne particulate matter and human health: toxicological assessment and importance of size and composition of particles for oxidative damage and carcinogenic mechanisms. *Journal of Environmental Science and Health, Part C: Environmental Carcinogenesis and Ecotoxicology Reviews*, 26, 339-362.

Veitch J, Salmon J, Carver A, et al. (2014) A natural experiment to examine the impact of park renewal on park-use and park-based physical activity in a disadvantaged neighbour-hood: the REVAMP study methods. *BMC Public Health*, 14, 1-9.

Viegi G, Simoni M, Scognamiglio A, et al. (2004) Indoor air pollution and airway disease. *International Journal of Tuberculosis and Lung Disease*, 8, 1401-1415.

Wang X Y, Barnett A G, Hu W, et al. (2009) Temperature variation and emergency hospital admissions for stroke in Brisbane, Australia, 1996-2005. *International Journal of Biometeorology*, 53, 535-541.

WAPC. (2009) *Liveable Neighbourhoods*: *A Western Australian Government Sustainable Cities Initiative*. Western Australian Planning Commission, Perth. http://www.planning.wa.gov.au/dop_pub_pdf/LN_Text_update_02.pdf (accessed 12/06/2015).

Weeks J R, Hill A, Stow D, et al. (2007) Can we spot a neighborhood from the air? Defining neighborhood structure in Accra, Ghana. *GeoJournal*, 69, 9-22.

Weiser S D, Leiter K, Bangsberg D R, et al. (2007) Food insufficiency is associated with high-risk sexual behaviour among women in Botswana and Swaziland. *PLoS Medicine*, 4, 1589-1598.

Weng Q, Yang S. (2004) Managing the adverse thermal effects of urban development in a densely populated Chinese city. *Journal of Environmental Management*, 70, 145-156.

White L A, Gehrt S D. (2009) Coyote attacks on humans in the United States and Canada. *Human Dimensions of Wildlife*, 14, 419-432.

World Bank. (2013) *Global Monitoring Report* 2013: *Rural-Urban Dynamics and the Millennium Development Goals*. World Bank, Washington DC. doi: 10.1596/978-0-8213-9806-7.

World Health Organization. (2011) *Intersectoral Actions in Response to Cholera in Zimbabwe*: *From Emergency Response to Institution Building*. World Conference on Social Determinants of Health, Rio de Janeiro, Brazil, Draft Background Paper no. 23. http://www.who.int/sdhconference/resources/draft_background_paper23_zimbabwe.pdf (accessed 12/06/2015).

World Health Organization, UN-Habitat. (2010) *Hidden Cities*: *Unmasking and Overcoming Health Inequities in Urban Settings*. WHO Press, Geneva.

Wyly E K, Hammel D J. (1999) Islands of decay in seas of renewal: housing policy and the resurgence of gentrification. *Housing Policy Debate*, 10, 711-771.

Zhang J, Smith K R. (2003) Indoor air pollution: a global health concern. *British Medical Bulletin*, 68, 209-225.

Zheng H W, Shen G Q, Wang H. (2014) A review of recent studies on sustainable urban renewal. *Habitat International*, 41, 272-279.

Zhou Z, Dionisio K L, Arku R E, et al. (2011) Household and community poverty, biomass use, and air pollution in Accra, Ghana. *Proceedings of the National Academy of Sciences of the United States of America*, 108, 11028-11033.

Zhu P, Zhang Y. (2008) Demand for urban forests in United States cities. *Landscape and Urban Planning*, 84, 293-300.

Zulu E M, Dodoo F N, Chika-Ezeh A. (2002) Sexual risk taking behaviour in the informal settlements of Nairobi, Kenya, 1993-1998. *Population Studies*, 56, 311-323.

第 7 章 保护城市中的生物多样性和维护生态系统服务

7.1 概述

如前几章所述，城市的建设和扩张可能会对本地生物多样性产生重大不利影响，并且当前世界各地的许多物种都受到城镇化的威胁（McKinney，2002；Pauchard et al.，2006；McDonald et al.，2008；Goddard et al.，2010）。然而，这不应使我们将城市视为不重要的生物保护场所。城市、郊区和近郊仍然能够支持众多本地物种和生态群落，包括从极度濒危的物种到目前最不受保护的物种（Yencken，Wilkinson，2000；Alvey，2006；Aronson et al.，2014a；Ives et al.，2016；专栏 7.1）。如果我们仅将保护工作集中在乡村和荒野地区，那么这些物种和群落中的许多就会消失。除了其内在的价值之外，城市中的生物多样性还对城市居民的健康和福祉具有重要的益处（详见第 6 章；Dearborn，Kark，2010；2012 年《生物多样性公约》）。来自城市生态学、人口健康、流行病学和环境心理学领域的大量知识表明，城市生物多样性增加了城市对人类的宜居性（Fuller et al.，2007；Tzoulas et al.，2007；Taylor，Hochuli，2015；Carrus et al.，2015）。

生态系统功能是由生态系统执行的功能，如养分循环、授粉、初级和二级生产、气候调节和污染物吸收（请参阅第 5 章）。生态系统服务是人类从生态系统功能中获得的益处，其中包括动植物食品、木材、植物纤维、清洁空气、清洁水和更舒适的气候（de Groot et al.，2012）。这些清单显然包括许多对人类生存至关重要的东西。全球生态系统服务的价值巨大，2011 年估计为每年 125 万亿美元（Costanza et al.，2014）。考虑到单个生境类型，珊瑚礁、湿地和热带森林提供的生态系统服务的价值,估计分别为每公顷每年 352249 美元、140174 美元和 5382 美元。在同样的分析中,2011 年城市栖息地提供的生态系统服务的价值为每公顷每年 6661 美元(Costanza et al.,2014)。城市生态系统提供的重要服务包括养分循环、遮光、降温、吸收大气污染物、固碳、初级和二级生产(包括都市农业)、水文循环、有害生物的防治和废物处理(Jo,2002;Li et al.,2005;Dwivedi et al.,2009)。众所周知,生物多样性可以增强城市的某些生态系统服务,包括养分循环、微气候降温、大气污染物的吸收和水文循环(Bolund,Hunhammer,1999;MEA,2005)。

在城市中,生物多样性的保护和生态系统服务的供给可能具有挑战性,在运作中存在许多相互竞争的利益(Eppink et al.,2004;Polasky et al.,2008;Gordon et al.,2009;UN-Habitat,2010;Bekessy et al.,2012)。由于土地经常短缺,价格高昂。这限制了新公园或保护区在城市内和城市附近购买土地的机会(Snyder et al.,2007),除了不适合其他类型开发的区域,如废弃的工业或垃圾填埋场、地形陡峭或容易遭受洪水泛滥的地区。较高的土地溢价还意味着城市剩余的生物多样性栖息地(如湿地、河岸走廊和残留的植被斑块),始终受到城市发展的威胁(Bekessy,Gordon,2007)。在许多情况下,短期的经济利益或其他社会经济方面的考虑,压倒了保护生物多样性和生态系统服务的长期利益(Crook,Clapp,1998;Balmford et al.,2002;Breed et al.,2014)。第三,城市不断向外扩张,这意味着目前离城市边缘一定距离的生物多样性区域,最终可能会被郊区或非正规居住区所覆盖(Polk et al.,2005;Benítez et al.,2012;Seto et al.,2012)。城市中的树木、草、动物、真菌、微生物、溪流、湿地和渗透性土壤所提供的生态系统服务通常未被人们认识或被低估(Taylor,Hochuli,2015),使得这些生物和栖息地元素易于被改变或破坏。本章介绍了一些在城市环境中保护生态多样性和提供生态系统服务的实用策略,以造福人类和生活在城市中的所有其他物种。

💬 专栏7.1

城市避让者、适应者和剥削者

McKinney(2002)首先使用"城市避让者、适应者和剥削者"这些术语,来描述以不同方式对城镇化做出反应的三类物种。他将城市避让者定义为对人类的迫害和栖息地干扰非常敏感的物种，例如大型食肉动物、地面筑巢的鸟类和演替后期植物。他认为城市适应者是可以利用现有的生态位和资源，在郊区和城市边缘栖息地成功生存的物种。这个群体包括杂食性和地面觅食的鸟类、挖洞的哺乳动物、中型捕食者、演替早期植物、靠鸟类传播的灌木（McKinney，2002）。城市剥削者也称为共生种，是一小类物种，对在城市环境中人类提供的资源表现出积极反应，并在世界各地的城市中获得很高的存活密度。该组动物包括岩鸽（*Columba livia*）、欧洲八哥（*Sturmus vulgarus*）、小家鼠（*Mus musculus*）、黑家鼠（*Rattus rattus*）、蟑螂（各种物种），靠种子传播和高抗干扰性的多种杂草（McKinney，2002）。

尽管McKinney（2002）的分类是有用的，但是术语"避让者"和"适应者""剥削者"暗示代表这些物种的特定意图或作用，这可能不合适。例如，城市避让者可能不会主动选择避开城市栖息地，取而代之的是，由于各种其他原因，这些物种的种群不能生存在城市环境中，例如，由于它们具有特殊的特征，使其容易受到城市干扰或掠夺，或者在城市景观中难以满足其资源需求。同样，城市适应者不一定会改变其行为或资源需求以适应城市环境中可用的资源。Grant等（2011）根据其对城镇化的反应确定了三组两栖动物和爬行动物——亲城市者、恐城市者（继Mitchell et al.，2008）和城市遗忘者。前两组大致对应McKinney定义的城市剥削者和城市避让者。第三类是由隐蔽的物种组成，它们可以在城市发展所包围、孤立的小片栖息地中生存多年，而这在很大程度上未被人类注意到。北美城镇遗忘者的例子包括长寿的大型物种[如鳄龟（*Chelydra serpentina*）]以及短命的小物种[如红背蝾螈（*Plethodon cinereus*]（Grant et al.，2011）。

物种的特定生态和生命史特征与它们在城市栖息地的持久性或被排斥性明显相关。这些特征和普遍性目前在城市生态学研究中引起了相当大的兴趣（Williams et al.，2005；Croci et al.，2008；Knapp et al.，2008；Hahs et al.，2009；Grant et al.，2011；见第4章的讨论）。有人提出，与城市避让者相比，对城市生活具有耐受性的物种在生态位方面具有更大的优势，这使它们在行为、生理和生态方面具有更大的灵活性，可以在变化的城市环境中坚持生存(Bonier et al., 2007)。这种模式似乎适用于鸟类,在城市环境中常见的物种比乡村环境中更丰富的同类物种,具有更广泛的纬度和海拔分布(Bonier et al., 2007)。它似乎也适用于两栖动物和爬行动物,在城市中常见的许多物种显示出广泛的栖息地和饮食耐受性(Mitchell et al., 2008; Grant et al., 2011)。在城市中可能无法对某些城市避让者/恐城市者进行保护,但是在其他情况下,有针对性的保护行动可能会使其中一些物种与人类共存。

7.2 城市生物多样性保护和维持生态系统服务的策略

7.2.1 将城市生态学与城市规划设计相结合

有效保护城市环境中的生物多样性和生态系统服务,需要各学科的投入,包括生态学、城市规划、风景园林学、生态经济学、工程学和景观管理(McPherson et al., 1997; Ong, 2003; Hunter, Hunter, 2008; Ahern, 2013; Tanner et al., 2014)。规划师、设计师和风景园林师在保护和促进城市中的生物多样性、绿色基础设施和生态系统服务方面发挥着重要作用,因为他们的实践对城市形态具有直接和持久的影响(Breed et al., 2014; Hes, Du Plessis, 2015)。通过将城市生态学更好地整合到各个空间尺度(这些空间尺度从个人住宅到社区、辖区和整个区域)的城市规划和景观管理中,得到许多机会来改善城市的生物多样性保护和生态系统服务(Jim, Chen, 2003; Uy, Nakagoshi, 2008; Gordon et al., 2009; Cilliers et al., 2014; Garrard, Bekessy, 2014)。

通过对城市发展相关的环境影响进行战略评估,可以将保护规划纳入大规模的城市规划。战略环境评估是对区域范围的评估,评估拟议发展对环境重要事项的影响包括受威胁物种、受威胁生态群落和具有生物学、社会和/或经济意义的自然地点等。战略评估可以在早期将环境价值和目标纳入大规模的政策和规划中,考虑替代发展方案,并在景观尺度上评价环境影响(Chaker et al., 2006)。与较小规模的环境影响评估相比,战略性环境评估可能为城市中的生物多样性和生态系统服务提供更好的保护，小规模环境影响评估通常缺乏长远的愿景、大范围计划和机制来考虑许多小规模行动的累积影响。这些累积影响被描述为"千刀万剐"；每个拟议的开发项目，都被视为与该区域的其他开发项目无关，并且其出现的环境影响还不足以阻止拟议的工程。但是，综合考虑起来，这些小行动对受威胁物种和生态群落具有持久性、重大的影响。另一方面，如果战略环境评估不够严格，则其区域规模方法可能通过大规模和快速清除栖息地的方式，这会加速城市区域本土物种和生态群落的丧失（Parris，2009）。

例如，澳大利亚州和地区政府已经根据1999年《联邦环境保护和生物多样性保护法案》编制了战略性环境评估，以获得联邦政府对墨尔本、悉尼和堪培拉拟议的城市发展计划的批准（DSE，2009；NSW Department of Planning，2010；Umwelt，2013）。2009年，墨尔本拟议的城市增长边界扩建，包括$4\times10^4 hm^2$的新住房、一条区域性铁路连接线和一条环城公路，将在城市周围的四条新增廊道内建造。这些廊道与联邦列出的濒危物种种群栖息的区域相吻合，如咆哮草蛙（*Litoria raniformis*）（图7.1）、条纹无腿蜥蜴、黄金蛾、南部棕色小袋鼠、多刺稻草花和亚麻百合以及极度濒危的植物群落残余（DSE，2009）。战略影响评估报告草案表明，拟议的开发项目将破坏大约$8000 hm^2$的极度濒危的天然温带草地、长满草的桉树林地和低地平原多草湿地，并对一系列受威胁物种产生重大（但未量化）的影响（DSE，2009）。另外，报告草案仅提出一种发展选择，并未评估拟议的开发项目对景观功能和连通性的影响。因此，拟议的开发项目不符合进行强有力的战略影响评估所需的所有标准（IAIA，2002；Chaker et al.，2006）。该拟议的开发项目的后续替代方案将濒临灭绝的植被群落的清除面积减少至约$4500 hm^2$。

维多利亚州政府正在通过强制性收购私有土地，在墨尔本西部建立两个新的草地保护区，以弥补因城市发展而流失的草地（DSE，2011）。作为战略影响评估流程的一部分，它为受威胁物种（包括咆哮草蛙）制定了生物多样性保护战略和一系列次区域物种策略（DEPI，2013a、2013b）。咆哮草蛙曾经在澳大利亚东南部广泛分布（Pyke，2002），但是自20世纪70年代后期以来，种群数量减少和局部灭绝意味着它现在的栖息地只占其原始范围的一小部分（Mahony，1999）。墨尔本周围的种群受到城市扩张造成的栖息地丧失、退化和破碎化的威胁（Heard et

图 7.1 一只咆哮草蛙（*Litoria raniformis*），该物种受到澳大利亚墨尔本市持续扩张的威胁（摄影：Geoff Heard）

al.，2012）；目前，墨尔本地区的咆哮草蛙的分布与城市北部、西部和东南部的新增廊道基本重合（DEPI，2013b）。然而，在城市生态与城市规划相结合的一个实际例子中，目前在使用咆哮草蛙的随机复合种群模型，来预测不同发展和管理场景对此物种的影响（Heard et al.，2013）。这些方案包括保护现有湿地，沿溪流保留各种宽度的缓冲区以及建立新的湿地（Heard et al.，2013）。该模型可以确定最佳解决方案和自然保护区的最佳边界，以便在给定的预算下最好地保护咆哮草蛙。

7.2.2 保护生物多样性景观特征和重要的生物物理资产

残余原生植被、次生植被、短暂和永久的湿地、溪流和排水管线的斑块，通常支持在城市景观中其他地方没有的高度多样性物种，同时还提供重要的生态系统服务，包括养分和水文循环、授粉、固碳和防洪减灾（Dearborn，Kark，2010；

Threlfall et al.，2012；Calhoun et al.，2014；Diaz-Porras et al.，2014）。尽管这些斑块的面积可能很小，这些特征的生态价值却很大；因此，它们有时被称为关键特征和结构（Stagoll et al.，2012；Calhoun et al.，2014）。因此，保留这些特征将为单位面积的城市生物多样性带来巨大的好处。例如，大而古老的树木充当城市环境中的关键结构，支持无脊椎动物、鸟类、蝙蝠和树栖哺乳动物的高度多样性，同时提供遮阴、微气候降温、固碳和重要的栖息地，如树洞和粗糙的木屑（Carpaneto et al.，2010；Lindenmayer et al.，2012；Stagoll et al.，2012；Diaz-Porras et al.，2014）。

在最近的一项研究中发现，随着澳大利亚堪培拉城市公园中大型树木的增加，鸟类的物种丰富度、多度和繁殖活动也随之增加；最大的树木（DBH＞100cm，胸径）表现出最大的栖息地保护效益（Stagoll et al.，2012）。在意大利罗马的比尔盖塞公园中，高大、活的空心树中，更容易出现腐生圣甲虫，包括自然保护联盟（IUCN）列出的隐士甲虫（*Qsmoderma eremita*），特别是那些树洞中有丰富木霉菌的树木（Carpaneto et al.，2010）。然而，这些树木中有许多被认为对人类构成危险，并已计划清除，这突出了生物多样性保护与城市公共安全问题之间的冲突（Carpaneto et al.，2010；Le Roux et al.，2014）。类似的冲突导致倒木稀缺，而倒木是城市环境中许多物种的另一个重要栖息地要素。在这种情况下，倒木被认为是危险的，因为人们普遍认为（在很大程度上不支持）倒木会为野火提供燃料（Lehvävirta，2007；Le Roux et al.，2014）。城市环境中，大型古树对人类构成危险，可以在古树周围设围栏或景观美化，以将其与公园高使用区域分开（Stagoll et al.，2012）。向大众宣传大直径倒木可能造成的实际火灾风险，这可能会增加它们在公园中的社会接受度。

在城市环境中保留高地排水管线（缺暂的一级河流），允许降雨后自然排水，并有助于减少进入河流和其他受纳水域的雨水量（Levick et al.，2008）。围绕排水管线设置线状公园，还可以为人类提供开放的绿色空间，为城市生物多样性提供栖息地（Uy，Nakagoshi，2008；Maric et al.，2013）。但是，由于这些排水管线仅在发生明显的降雨之后才有水流，因此它们通常会被盖起来，其自然排水功能被工程结构代替。保留城市排水管线的两个成功实例是加利福尼亚州戴维斯的乡村住宅区的自然排水系统——小溪、沼泽和池塘组成的网络，它可以保留雨水并在土壤中储存水分（Corbett，Corbett，2000；Karvonen，2011）；类似的例子，还有德国汉诺威附近的克朗斯堡（Kronsberg）社区（Rumming，2004；Coates，2013）。这类设计实践可以减少城市集水区和下游水域的退化，为人类娱乐休闲和栖息地的连通性提供绿色空间，并在更广泛的城市景观中促进生物多样性（Urbonas，Doerfer，2005；Walsh et al.，2005；von Haaren，Reich，2006）。

城市中的小片残余植被仍可以继续支持令人惊讶的本地物种的高度多样性，尽管城市发展已将它们包围起来（Williams et al.，2006；Gupta et al.，2008；Newbound et al.，2012）。例如，一项对澳大利亚墨尔本残余林地斑块中的真菌调查发现，分析土壤样本中真菌的 DNA，发现每个斑块中有 54～114 种真菌（Newbound et al.，2012）；而至少有 135 种维管束植物种群仍然生存在印度那格浦尔的城市森林残余斑块中（Gupta et al.，2008）。半结构化的栖息地，例如高尔夫球场和休闲公园，也可以成为城市生物多样性的重要空间和生态系统服务的重要提供者，尤其是在保留某些自然植被或地面覆盖物的情况下，如落叶和未修剪的草（Savard et al.，2000；Barthel et al.，2005；Tyrvainen et al.，2005；Yasuda，Koike，2006；Colding，Folke，2008；Stagoll et al.，2012）。对澳大利亚墨尔本的高尔夫球场、城市公园和住宅花园的一项研究发现，高尔夫球场支持的甲虫、本地蜜蜂和昆虫（异翅目）的多样性和多度比城市附近的公园和花园更大，包括对控制虫害很重要的捕食性昆虫（Threlfall et al.，2014）。高尔夫球场还容纳更多种类的鸟类和蝙蝠，相比于其他两种栖息地类型，几乎支持两倍的鸟类繁殖活动。随着周围土地受到更大的人为影响，高尔夫球场作为生物多样性避风港和生态系统服务提供者的重要性可能会增加（Colding，Folke，2008）。旧的砾石坑是另一种半结构化的栖息地，对于保护城市环境中的无脊椎动物（包括蝴蝶）可能很重要（Lenda et al.，2012）。

私人花园和行道树是城市中广泛分布的活资产，在城市景观中占树木和可渗透土壤总覆盖量的很大一部分（Cameron et al.，2012）。因此，它们在提供生态系统服务方面发挥着重要作用，如遮阴、微气候降温、养分循环、初级生产、固碳、降噪、吸收大气污染物、水文循环以及与自然的连通感（Livesley et al.，2010；Kendal et al.，2012；Norton et al.，2013；Clarke et al.，2014；Stovin et al.，2015）。私人花园还支持许多分类群中相当大比例的城市生物多样性，从土

7.2 城市生物多样性保护和维持生态系统服务的策略

壤微生物和真菌,到观赏、食用和药用植物,虫子、蜜蜂、蝴蝶、爬行动物和鸟类(Gaston et al.,2005;Loram et al.,2008;Akinnifesi et al.,2010;Goddard et al.,2010;Kendal et al.,2010;Jaganmohan et al.,2012;Cilliers,Siebert,2012)。然而,城市填空建设和建造大型房屋的趋势意味着许多城市私人花园的覆盖比例(以及它们所支持的生态系统服务和生物多样性)正在下降(Marriage,2010;Vallance et al.,2012;Brunner,Cozen,2012)。

在世界某些地区有一种趋势是,大型房屋所占比例比几十年前高得多,而可用作花园的区域却很小(Marriage,2010;Hall,2011)。2009 年,澳大利亚拥有世界上最大的独立式新房,平均建筑面积为 245.3m^2,20 年来增长了 34%(James,2009)。根据美国人口普查数据,2013 年美国新建单户住宅的平均建筑面积为 241.4m^2(图 7.2;US Census Bureau,2014)。在新西兰也可以看到类似的趋势,2011 年新房屋的平均建筑面积为 219m^2,而 1991 年新房屋的平均建筑面积为 139m^2(Marriage,2010;图 7.2)。房屋与花园的比值高,降低了舒适性;失去了树木遮阴和绿色植物蒸腾所带来的微气候降温的好处,吸热不透水表面的覆盖率增加,户外休闲活动的空间也减少了。这些现象使当地气候变得更热和更干

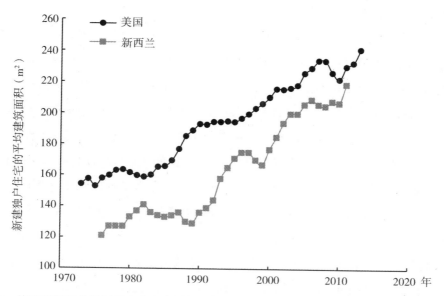

图 7.2 美国和新西兰每年新建独户房屋的平均建筑面积的变化情况;在过去的几十年中,这两个国家的房屋面积都大幅增加(数据来源:Marriage,2010;US Census Bureau,2014)

燥，这种城市形式增加了夏季空调的使用和整体能源消耗(在许多情况下，还增加了化石燃料的燃烧)。它还减少了土壤吸收的雨水量，增加了雨水流入河流的流量(Brunner，Cozen，2012)。

7.2.3 发展绿色城市

考虑到行道树和花园对于生物多样性和城市宜居性至关重要，而且在气候变化的影响下可能变得更加重要(Gago et al.，2013；Norton et al.，2013)，城市需要有远见的城市绿化政策来增加林木覆盖率，并需要加强监管以保护现有的植被和渗透性土壤。这些法规已经在许多城市中实施，包括柏林("生物区面积系统")和西雅图("西雅图绿色因子")。得克萨斯州伍德兰兹的一项树木保护政策规定了乔木和灌木的最小覆盖率，并严格监管树木砍伐，要求砍伐任何直径≥15cm 的树木必须得到许可(Sung，2013)。这项政策有效地减少了城市热岛效应，其平均地表温度比附近没有树木保护政策的对照区低 1.5~3.9℃。夏季最需要散热，伍德兰兹镇因有大量植被，降温效应更显著，平均温差最高为 5℃，观测到的最大温差为 7.1℃(Sung，2013)。在本研究的 7 个社区中，5 月炎热天气中，地表温度随着林木覆盖率的增加而降低(图 7.3)。

如第 6 章所述，各种研究表明城市树木和其他多叶植被的覆盖率与社区的社会经济状况呈正相关；在许多城市中，较富裕地区的树木更多(Pedlowski et al.，2002；Escobedo，2006；Harlan et al.，2007；Weeks et al.，2007；Clarke et al.，2013；Shanahan et al.，2014；Hetrick et al.，2013)。城市树木和其他植被在城市之间分布不均，这意味着城市树木覆盖所带来的社区利益和生态系统服务也分布不均。因此，城市树木覆盖被认为是社会公正问题(Wolch et al.，2014)。假设林木覆盖率降低 5%，可以使气温升高 1~2℃，那么富裕区和贫穷郊区之间的林木覆盖率差异可能会导致树木最多和最少的地区之间的温度差异很大(McPherson，Rowntree，1993)。为了提高城市树木覆盖的公平性，城市绿化政策和植树政策应针对植被覆盖率低且易受极端高温影响的居住社区(Norton et al.，2015)。这些策略可以包括鼓励家庭住户在自己花园中增加树木和灌木植物覆盖率的激励计划，以及在街道和公园种植更多的树木(Shanahan et al.，2014)。在炎热和干燥的气候条件下，储存的雨水可用于灌溉公园、花园和行道树，以提高城市的凉爽程度，并改善人体热舒适度（Coutts et al.，2013）。这将带来减少雨水流入当地溪流和其他受纳水域的额外好处。

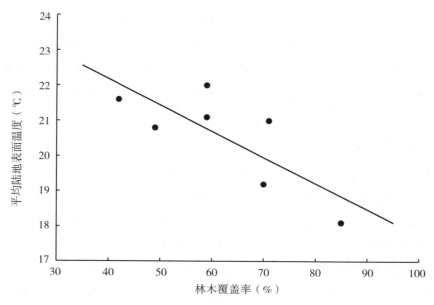

图 7.3　城市树木覆盖的降温效果。美国得克萨斯州休斯顿市区北部 7 个社区的平均地表温度与林木覆盖率之间的对比。温度数据来自 2000 年至 2010 年期间拍摄的 37 幅 Landsat TM 热红外遥感图像；树木覆盖数据来自美国国家农业影像计划（NAIP），航空摄影时间为 2010 年 5 月 3 日（Sung，2013）

7.2.4　维护或重建景观连通性

城市中的栖息地特征，如公园、湿地以及残余或次生的植被斑块，往往被宽阔的道路、建筑物和其他城市基础设施彼此隔开。如第 3 章所述，这可能会阻止或妨碍繁殖体和个体在离散的栖息地之间成功移动，从而有效地隔离了许多真菌、动植物的种群（Beier，1995；Bjurlin et al.，2005；Parris，2006；Fu et al.，2010；Hale et al.，2013；Ward et al.，2015）。为了避免这种隔离，在城市规划过程中需要考虑景观连通性，最好是在开发新城市或郊区之前，保留未开发的线性区域以充当生物多样性景观特征之间的廊道或管道，将有助于维持城市环境中某些生物分类群体(包括节肢动物、两栖动物和哺乳动物)之间的连通性(Beier, Noss, 1998; Indykiewicz et al., 2011; Vergnes et al., 2012, 2013; Hale et al., 2013; Wilson et al., 2013)。在法国巴黎,节肢动物利用树木繁茂的廊道穿越城市景观。这些廊道连接私人花园和林地,被发现栖息着节肢动物群落,其功能与林地相似(Vergnes et al., 2012)。与不相连的花园相比,相连的花园生活着更多的蜘蛛和甲虫。另一项研究发现,这些相同的廊道被鼩鼱用于从林地移动到私家花园,43%的连通花园中观察到鼩鼱,仅有6%的不连通花园中观察到鼩鼱(Vergnes et al., 2013)。

越南河内的城市绿地规划包括公园、大型楔状绿地、条状绿化带以及沿路旁和溪流的许多其他线性绿地,共同构成了生态有效的绿色网络(图7.4; Uy, Nakagoshi, 2008)。该网络旨在连接不同的绿色空间,并允许动植物在它们之间移动。绿色网络包含现有的绿色空间,以及未来将要建设的其他绿道,并将自然的和已构建的线性绿地合并到绿色网络中。河内绿地规划表明,改造既有城市区域以增强栖息地连通性的可能性;然而,与建立新的景观连接相比,保留现有的景观连接更容易,并且可能更经济(Benedict, McMahon, 2006)。鉴于栖息地廊道和其他景观连接占据了一定的物理空间,因此在人口密集的城市环境中重建旧廊道或建立新廊道的计划可能会受到限制。如果满足某些条件,地下通道、立交桥、野生动物桥梁和其他允许动物横穿的道路工程设施,也可能对提供城市栖息地连通性作出宝贵贡献;目标动物必须能够穿越数量较多的设施,并且要确保免受捕食者的攻击,保护自身安全(Ng et al., 2004; van der Grift et al., 2013; Hamer et al., 2014; Ward et al., 2015)。

人类可以通过将单个植物或动物从一个位置移动(或迁移)到另一个位置,或通过帮助迁徙动物安全地到达目的地来协助动植物在整个城市景观中的移动。虽然迁移通常被认为是城市栖息地物种保护的合适工具,如将个体从一个地方转移到另一个显然合适的地点,但是通常成功的可能性很小。对爬行动物、两栖动物和哺乳动物的许多迁移由于各种原因而失败,包括释放动物的归巢、放生的地点缺乏合适的、无人居住的栖息地,或随着时间的推移,栖息地的适宜性下降,释放个体数量不足、被捕食或疾病(Pietsch, 1994; Fischer, Lindenmayer, 2000; Armstrong, 2008; Germano, Bishop, 2009; Koehler, Gilmore, 2015)。尽管从直

7.2 城市生物多样性保护和维持生态系统服务的策略

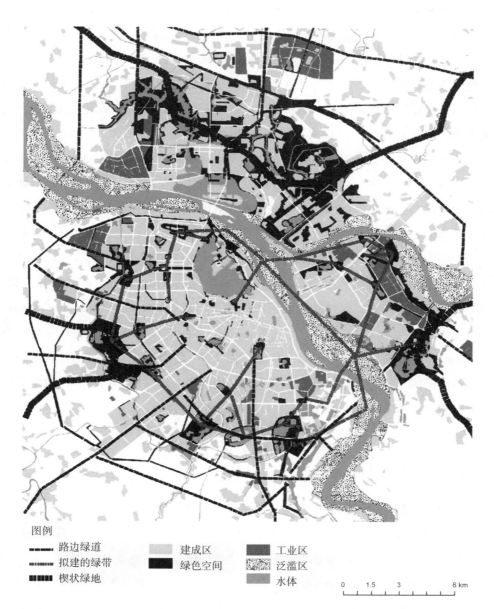

图 7.4 越南河内的城市绿地规划，其中包括公园、楔状绿地和沿路边、溪流形成的线性绿地构成的生态有效网络（Uy，Nakagoshi，2008，图10）

观上来看很吸引人，但是迁移并不是拯救城市景观中重要栖息地特征被破坏的灵丹妙药，如湿地或残余植被斑块。然而，在某些情况下，迁移可能有助于在城市过于孤立而无法自然地定居的物种在合适的、空置的栖息地重新定居（Martell et al.，2002；Parris，2006；White，Pyke，2008）。

季节性地从栖息地的一部分迁移到另一部分的动物，往往年复一年地沿着相同的路线到达相同的目的地，而忽略了栖息地景观的变化。例如，青蛙可能达到一个不再存在的湿地繁殖（Windmiller，Calhoun，2008）。每年成千上万只青蛙、火蜥蜴、蝾螈、乌龟，试图穿越去往繁殖场地的道路而死亡（Gibbs，Shriver，2002，2005；Marchand，Litvaitis，2004；Windmiller，Calhoun，2008）。包括南非、英国、荷兰、罗马尼亚、德国、丹麦和爱沙尼亚在内的许多国家、社区团体已经动员起来，通过一种称为巡逻的方式，防止两栖动物在道路上被杀伤。志愿者在适当的季节巡逻重要的迁徙道路口，并帮助动物安全地过马路。"青蛙生活"社团在英国实施针对普通蟾蜍（*Bufo bufo*）的"蟾蜍在路上"计划，允许社团成员注册一个蟾蜍过境点并自愿作为蟾蜍巡逻者（www.froglife.org）。南非开普敦，也开展了类似的计划，以帮助濒临灭绝的西部豹蟾蜍（*Amietophrynus pantherinus*）安全地进出繁殖地。"Toad NUTS"（Noorhoek 公益性蟾蜍保护者）小组最近在 Noorhoek（开普敦的一个郊区）的一条繁忙的道路上建立了带有陷阱的漂流栅栏，以拦截移动的蟾蜍，将它们安全地带到另一边。还有在两栖动物频繁出没的道路两旁设置限速和警告驾驶者的路标，甚至季节性封路以保护迁徙动物（Hilty et al.，2006；Wang et al.，2013；图 7.5）。

7.2.5 利用小空间

城市环境可能会挤满人造建筑、铺砌路面以及精细拼接的土地利用。这可能很难找到自然区域以及植被、渗透性土壤、湿地和其他景观特征的空间，而这些景观空间是生物多样性和生态系统服务提供者的栖息地。解决此问题的一种方法是在整个城市中使用较小的空间：水平、垂直以及介于两者之间的所有空间（Rosenzweig，2003a，2003b）。调和生态学（Reconciliation Ecology）是生态学的一个分支，强调在人类主导的景观中保护生物多样性。调和生态学于 2003 年首次引入，可以定义为"发现、建立和维护新的栖息地，以保护人们生活、工作或娱乐场所的物种多样性的科学"（Rosenzweig，2003a）。调和生态学的重点之一是城市环境的小规模或局部变化，可以帮助保护和增强公共、私人土地上的生物多样性。野生动植物花园的做法是调和生态学的一种形式,也是通过许多小型私人花园是增加城市生物多样性的一个有效途径(Rudd et al., 2002; Colding, 2007; Gaston et al., 2007; Goddard et al., 2013)。

图 7.5 德国 Angelbachtal 的一个标志,警告驾驶者青蛙可能正在横穿过马路(摄影:Jencu)

一个大城市可能包含成千上万个平屋顶建筑,可以将其改造成活动的绿色屋顶。绿色屋顶可以提供一系列益处,包括对建筑物进行隔热和热调节,通过生长介质吸收和植物蒸散来循环雨水,为各种生物类群提供栖息地,吸收城市大气污染物,并且人们在一天的工作中抬头看到绿色屋顶,可以提高生产力和维持精神健康(Kadas, 2002; Wong et al., 2003; Mentens et al., 2006; Oberndorfer et al., 2007; Currie, Bass, 2008; Castleton et al., 2010; Braaker et al., 2014; Lee et al., 2015)。建造屋顶花园,可以提供一个放松、愉快的空间,另外种植草药、水果和蔬菜,甚至可以养蜜蜂酿蜜;这样,它们提供食物、舒适性,与自然联系,以及包括授粉在内的各种其他生态系统服务(Shariful Islam, 2004; Wong et al., 2003; Grewal, Grewal, 2012)。在世界许多城市中,城市屋顶养蜂业都在蓬勃发展;伦敦以3000多个屋顶蜂箱为首,而在上海、檀香山、布里斯班和纽约也有屋顶养蜂(Saggin, 2013)。蜜蜂在距其蜂巢小于5km范围内觅食,因此可以为整个城市景观中的多种园林植物授粉。绿色屋顶也可以为本土蜂种提供栖息地,尽管它们比蜜蜂更神秘,其本身也是重要的授粉媒介(Oberndorfer et al., 2007; Colla et al., 2009; Tonietto et al., 2011)。

绿墙或垂直绿化是植物覆盖在城市垂直表面。它们由攀援植物组成,直接在建筑物外墙上生长,或附在墙壁支撑结构上(包括棚架、钢缆、铁丝网)。垂直绿化在城市环境中提供了许多好处,包括遮阳、降温(通过蒸散)、提高太阳辐射的反射率、降低墙体表面的风速、吸收大气污染物和提供生物多样性的栖息地(Currie, Bass, 2008; Köhler et al., 2008; Hunter et al., 2014)。使用支撑结构的绿色立面,还会在这些结构和墙壁之间形成一层空气,从而使墙壁及其建筑物在夏季和冬季均不受极端温度的影响(Hunter et al., 2014)。

绿色立面在狭窄的城市街道(也称为城市峡谷)中特别有用,因为在狭窄的城市街道上,植树空间很小(Norton et al., 2015)。在炎热干燥的气候下,将攀援植物种植在朝西的墙壁上,可以提供最大的热效益(Hunter et al., 2014)。不过,城市墙壁上的微气候条件(如温度和风速)可能非常极端,因此需要细心选择植物、生长介质和灌溉系统(在某种情况下),以确保绿色外墙蓬勃发展。植物还可以自发地在城市中人工建造的墙壁上建植,有时形成分类和结构上的多样化群落,并支持稀有或受威胁的物种(Francis, 2011)。例如,在香港以绞杀榕为主的植物群落在石质挡土墙上生长,为原本单调的城市景观提供了绿色(Jim, 2014)。食果鸟类和蝙蝠带来的绞杀榕种子在挡土墙砌块之间的接缝处发芽,随着绞杀榕的生长,它们会抓住墙壁(充当替代宿主树)。

7.3 新型的栖息地和生态系统

城镇化为动物和植物营造了一系列新型的栖息地，其中包括道路廊道、铁路、建筑、桥梁、休闲公园、私人花园、人工湿地、花园池塘、人工海堤和其他海洋结构（如码头和浮桥）。这些栖息地为生物多样性提供了重要资源：土壤、养分、食物、水、庇护所、筑巢、栖息和繁殖场所或定居的基质。对于某些可以利用这些新颖栖息地的物种，它们可以部分补偿城镇化后自然栖息地的丧失。例如，许多海洋植物和动物可以成功地在海堤上定居，包括藻类、固着无脊椎动物（如管虫、藤壶、贻贝）和可移动无脊椎动物（如帽贝、螃蟹和蜗牛）（Chapman, Blockley, 2009; Ng et al., 2012）。在悉尼港进行的一项研究发现，尽管与自然的、缓坡的潮间带海岸相比，垂直潮间带海堤通常能支撑较少的物种，尤其是可移动无脊椎动物更少，但是海堤仍然为100多个物种提供了栖息地（Chapman, 2003）。城市区域的海洋结构物还可以充当鱼类的栖息地，在某些情况下还可以充当人工鱼礁。在以色列埃拉特的红海中，支撑着两个石油码头的支柱吸引了35个科的146种鱼类，其中包括当地鱼类、属地（地盘性）鱼类和到访鱼类（Rilov, Benayahu, 1998）。与附近退化的珊瑚礁相比，这种多样性具有优势。码头的支柱是钢制的，部分被铁丝网包围；金属丝在柱子上的覆盖范围越大，鱼的多度和多样性就越大。铁丝网和柱子之间为幼鱼提供了庇护所，降低了捕食风险（Rilov, Benayahu, 1998）。

西印度群岛的各种本土和引进的爬行动物物种，定居在建筑物内或建筑物上（也被称为建筑栖息地）或成堆的城市垃圾上，这些地方提供了庇护所以及大量节肢动物猎物（Powell, Henderson, 2008）。尼加拉瓜莱昂的私人花园为黑棘尾鬣蜥（*Ctenosaura similis*）提供了重要的栖息地（Gonzalez-Garcia et al., 2009；图7.6）。这种鬣蜥被称为地球上最快的蜥蜴，其时速可以达到34.6km/h（Garland, 1984），雄性体长可达1.5m。莱昂建在一个历史上曾经为热带干旱森林的地区，其林木覆盖率高于周围的农业用地，庭院绿地占城市绿地总面积的85%以上。人们发现具有高头树木、鬣蜥喜欢的食用植物和可穿透的栅栏的大型庭院，可以支持最高的黑棘尾鬣蜥多度（Gonzalez-Garcia et al., 2009）。成年鬣蜥大部分时间都待在树上或房屋的屋顶上，也许是为了躲避捕猎它们的人类。被调查的1/4的庭院中有鬣蜥洞穴，其中一半建在成堆的垃圾和建筑材料中。这几个例子说明了海洋和陆地环境中城市居住物种如何开发新型的栖息地。

图7.6 尼加拉瓜莱昂的私人花园中可以发现黑棘尾鬣蜥（摄影：Christian Mehlführer）

新型生态系统(有时称为新兴生态系统)的特征是存在新的物种组合，物种新的相对多度以及人类行为可能导致的生态系统功能变化(Chapin, Starfield, 1997; Milton, 2003; Hobbs et al., 2006)。它们被认为是新颖的，因为在历史上没有类似物(Seastedt et al., 2008; Kowarik, 2011; Hobbs et al., 2013; Graham et al., 2014)。新型生态系统(包括新的城市生态系统)，可能很难或不可能恢复到以前的状态，因此需要相应新颖的管理方法，来最大化其生物多样性和其提供的生态系统服务，而不是花费资源试图重建过去的生态系统(Hobbs et al., 2006, 2009, 2013; Seastedt et al., 2008)。新型生态系统的概念在恢复生态学领域引起了一定的争议，一些作者认为这对生态系统的定义不明确，令人困惑，可能对保护造成破坏（Aronson et al.，2014b；Murcia et al.，2014）。有人担心，将某些生态系统归类为新型生态系统可能会导致人们放弃对其进行恢复，而实际上这些生态系统的重要部分是可以挽救的（Aronson et al.，2014b；Murcia et al.，2014）。当然，尽管新型城市生态系统的状态发生了变化（包括非本土物种的存在），但它们仍具有生态价值，并且通过适当的管理，有可能提供进一步的生态价值，这一观点是强有力的（Standish et al.，2013）。这一价值可能表现为生物多样性的栖息地、提供生态系统服务或提供城市居民与自然互动的场所和空间。

7.4 总结

在城市中保护生物多样性和提供生态系统服务并非没有挑战，但是这些双重努力对人类和生活在城市中的其他物种的好处是巨大的。加强城市生物多样性和生态系统服务的重要战略，包括将城市生态与城市规划设计相结合；保护重要的生态多样性栖息地和景观特征；增加城市树木和其他植被的总覆盖率，并增加其在城市中分布的公平性；维持或改善景观连通性；以及将小型城市空间改造成自然空间。尽管城市生态系统的状态已发生变化，认清城市生态系统的现在价值和未来潜力，将使我们能够保护更广泛的物种多样性，并从生态系统服务中获益，包括遮阴、降温、改善大气污染、养分和水文循环以及粮食生产。随着世界气候的变化，这些服务不仅能改善我们城市的舒适性和宜居性，而且对许多城市栖居者来说，这可能意味着生与死之间的差异。

思考题

1. 为什么城市区域对于生物多样性保护很重要？
2. 考虑一个在你附近的城市或郊区濒临灭绝的物种和生态群落，它容易受到城市进一步扩张的影响吗？可以采取什么措施来保护它？
3. 讨论生态系统服务对于城市宜居的至关重要性。
4. 城市生态如何与城市规划设计相结合，用以改善生态多样性保护和城市生态系统服务的供给？
5. 解释以下各项在城市环境中的重要性：生物多样性景观特征、蒸腾植被和景观连通性。

6. 什么是城市适应者、避让者和剥削者？你如何识别每个组？

7. 城市生态系统是新型生态系统，讨论其原因。

本章参考文献

Ahern J. (2013) Urban landscape sustainability and resilience: the promise and challenges of integrating ecology with urban planning and design. *Landscape Ecology*, 28, 1203-1212.

Akinnifesi F K, Sileshi G W, Ajayi O C, et al. (2010) Biodiversity of the urban homegardens of São Luís city, Northeastern Brazil. *Urban Ecosystems*, 13, 129-146.

Alvey A A. (2006) Promoting and preserving biodiversity in the urban forest. *Urban Forestry and Urban Greening*, 5, 195-201.

Armstrong A J. (2008) Translocation of black-headed dwarf chameleons *Bradypodion melanocephalum* in Durban, KwaZulu-Natal, South Africa. *African Journal of Herpetology*, 57, 29-41.

Aronson M F J, La Sorte F A, Nilon C H, et al. (2014a) A global analysis of the impacts of urbanization on bird and plant diversity reveals key anthropogenic drivers. *Proceedings of the Royal Society B*, 281, 20133330. http://dx.doi.org/10.1098/rspb.2013.3330.

Aronson J, Murcia C, Katten G H, et al. (2014b) The road to confusion is paved with novel ecosystem labels: A reply to Hobbs et al. *Trends in Ecology and Evolution*, 29, 646-647.

Balmford A, Bruner A, Cooper P, et al. (2002) Economic reasons for conserving wild nature. *Science*, 297, 950-953.

Barthel S, Colding J, Elmqvist T, et al. (2005) History and local management of a biodiversity-rich, urban cultural landscape. *Ecology and Society* 10, 10. www.ecologyandsociety.org/vol10/iss2/art10/.

Beier P. (1995) Dispersal of juvenile cougars in fragmented habitat. *The Journal of Wildlife Management*, 59, 228-237.

Beier P, Noss, R. F. (1998) Do habitat corridors provide connectivity? *Conservation Biology*, 12, 1241-1252.

Bekessy S A, Gordon A. (2007) Nurturing nature in the city. In *Steering Sustainability in an Urbanizing World: Policy* (ed A. Nelson), Practice and Performance. Ashgate Publishing, Hampshire, 227-238.

Bekessy S A, White M, Gordon A, et al. (2012) Transparent planning for biodiversity and development in the urban fringe. *Landscape and Urban Planning*, 108, 140-149.

Benedict M A, McMahon E T. (2006) *Green Infrastructure: Smart Conservation for the 21st Century*. The Sprawl Watch Clearinghouse, Washington DC.

Benítez G, Perez-Vazquez A, Nava-Tablada M, et al. (2012) Urban expansion and the environmental effects of informal settlements on the outskirts of Xalapa city, Veracruz, Mexico. *Environment and Urbanization*, 24, 149-166.

Bjurlin C D, Cypher B L, Wingert C M, et al. (2005) *Urban Roads and the Endangered San Joaquin Kit Fox*. FHWA/CA/IR-2006/01. California Department of Transportation, Fresno. http://www.dot.ca.gov/newtech/researchreports/2002-2006/2005/urban_roads_and_the_endangered_san_joaquin_kit_fox.pdf (Accessed 15/02/2015).

Bolund P, Hunhammer S. (1999) Ecosystem services in urban areas. *Ecological Economics*, 29, 293-301.

Bonier F, Martin P R, Wingfield J C. (2007) Urban birds have broader environmental tolerance. *Biology Letters*, 3, 670-673.

Braaker S, Ghazoul J, Obrist M K, et al. (2014) Habitat connectivity shapes urban arthropod communities: the key role of green roofs. *Ecology*, 95, 1010-1021.

Breed C, Cilliers S, Fisher R. (2014) Role of landscape designers in promoting a balanced approach to green infrastructure. *Journal of Urban Planning and Development*, 141, doi: 10.1061/(ASCE)UP.1943-5444.0000248.

Brunner J, Cozens P. (2012) Where have all the trees gone? Urban consolidation and the demise of urban vegetation: a case study from Western Australia. *Planning*,

Practice and Research, 28, 231-255.

Calhoun A J K, Arrigoni J, Brooks R P, et al. (2014) Creating successful vernal pools: a literature review and advice for practitioners. *Wetlands*, 34, 1027-1038.

Cameron R W F, Blanusa T, Taylor J E, et al. (2012) The domestic garden-its contribution to urban green infrastructure. *Urban Forestry and Urban Greening*, 11, 129-137.

Carpaneto G M, Mazziotta A, Coletti G, et al. (2010) Conflict between insect conservation and public safety: the case study of a saproxylic beetle (*Osmoderma eremita*) in urban parks. *Journal of Insect Conservation*, 14, 555-565.

Carrus G, Scopelliti M, Lafortezza R, et al. (2015) Go greener, feel better? The positive effects of biodiversity on the well-being of individuals visiting urban and peri-urban green areas. *Landscape and Urban Planning*, 134, 221-228.

Castleton H F, Stovin V, Beck S B M, et al. (2010) Green roofs: building energy savings and the potential for retrofit. *Energy and Buildings*, 42, 1582-1591.

Chaker A, El-Fadl K, Chamas L, et al. (2006) A review of strategic environmental assessment in 12 selected counties. *Environmental Impacts Assessment Review*, 26, 15-56.

Chapin F S, Starfield A M. (1997) Time lags and novel ecosystems in response to transient climatic change in arctic Alaska. *Climate Change*, 35, 449-461.

Chapman M G. (2003) Paucity of mobile species on constructed seawalls: effects of urbanization on biodiversity. *Marine Ecology Progress Series*, 264, 21-29.

Chapman M G, Blockley D J. (2009) Engineering novel habitats on urban infrastructure to increase intertidal biodiversity. *Oecologia*, 161, 625-235.

Cilliers S, Siebert S J. (2012) Urban ecology in Cape Town: South African comparisons and reflections. *Ecology and Society* 17, 33. http://www.ecologyandsociety.org/vol17/iss3/art33/.

Cilliers S, du Toit M, Cilliers J, et al. (2014) Sustainable urban landscapes: South. African perspectives on transdisciplinary possibilities. *Landscape and Urban Planning*, 125, 260-270.

Clarke L W, Jenerette G D, Davila A. (2013) The luxury of vegetation and the legacy of tree biodiversity in Los Angeles, CA. *Landscape and Urban Planning*, 116, 48-59.

Clarke L W, Li L, Jenerette G D, et al. (2014) Drivers of plant biodiversity and ecosystem service production in home gardens across the Beijing Municipality of China. *Urban Ecosystems*, 17, 741-760.

Coates G. (2013) Sustainable urbanism: Creating resilient communities in the age of peak oil and climate destabilization. In *Environmental Policy is Social Policy—Social Policy is Environmental Policy*. Wallimann I ed. Springer, New York, 81-101.

Colding J. (2007) "Ecological land-use complementation" for building resilience in urban ecosystems. *Landscape and Urban Planning*, 81, 46-55.

Colding J, Folke C. (2009) The role of golf courses in biodiversity conservation and ecosystem management. *Ecosystems*, 12, 191-206.

Colla S R, Willis E, Packer L. (2009) Can green roofs provide habitat for urban bees (*Hymenoptera: Apidae*)? *Cities and the Environment*, 2, 1-12.

Corbett J, Corbett M. (2000) *Designing Sustainable Communities: Learning From Village Homes*. Island Press, Washington DC.

Costanza R, de Groot R, Sutton P, et al. (2014) Changes in the global value of ecosystem services. *Global Environmental Change*, 26, 152-158.

Coutts A M, Tapper N J, Beringer J, et al. (2013) Watering our cities. The capacity for Water Sensitive Urban Design to support urban cooling and improve human thermal comfort in the Australian context. *Progress in Physical Geography*, 37, 2-28.

Croci S, Butet A, Clergeau P. (2008) Does urbanization filter birds on the basis of their biological traits? *The Condor*, 110, 223-240.

Crook C, Clapp R A. (1998) Is market-oriented forest conservation a contradiction in terms? *Environmental Conservation*, 25, 131-145.

Currie B A, Bass B. (2008) Estimates of air pollution mitigation with green plants and green roofs using the UFORE model. *Urban Ecosystems*, 11, 409-422.

Dearborn D C, Kark S. (2010) Motivations for conserving urban biodiversity.

Conservation Biology, 24, 432-440.

de Groot R, Brander L, van der Ploeg S, et al. (2012) Global estimates of the value of ecosystems and their services in monetary units. *Ecosystem Services*, 1, 50-61.

DEPI. (2013a) *Sub-regional Species Strategy for the Growling Grass Frog*. Victorian Government Department of Environment and Primary Industries, Melbourne. http：//www. depi. vic. gov. au/__data/assets/pdf_file/0016/204343/GGF-SSS-text-only. pdf(accessed 25/06/2015).

DEPI. (2013b) *Biodiversity Conservation Strategy for Melbourne's Growth Corridors*. Victorian Government Department of Environmentand Primary Industries, Melbourne. http：//www. depi. vic. gov. au/__data/assets/pdf_file/0012/204330/BCS-no-maps-PtA. pdf(accessed 25/06/2015).

Diaz-Porras D F, Gaston K J, Evans K L. (2014) 110 years of change in urban tree stocks. and associated carbon storage. *Ecology and Evolution*, 4, 1413-1422.

DSE. (2009) *Delivering Melbourne's Newest Sustainable Communities：Strategic Impact Assessment Report for Environment Protection and Biodiversity Conservation Act* 1999. Victorian Government Department of Sustainability and Environment, Melbourne.

DSE. (2011) *Western Grassland Reserves：Grassland Management Targets and Adaptive Management* (2011). Victorian Government Department of Sustainability and Environment, Melbourne. http：//www. depi. vic. gov. au/__data/assets/pdf_file/0017/204371/WGR_TAM. pdf(accessed 25/06/2015).

Dwivedi P, Rathore C S, Dubey Y. (2009) Ecological benefits of urban forestry：The case of Kerwa Forest Area (KFA), Bhopal, India. *Applied Geography*, 29, 194-200.

Eppink F V, van den Bergh J C J M, Rietveld P. (2004) Modelling biodiversity and land use：urban growth, agriculture and nature in a wetland area. *Ecological Economics*, 51, 201-216.

Escobedo F J, Nowak D J, Wagner J E, et al. (2006) The socioeconomics and management of Santiago de Chile's public urban forests. *Urban Forestry and Urban Greening*, 4, 105-114.

Fischer J, Lindenmayer D B. (2000) An assessment of the published results of animal relocations. *Biological Conservation*, 96, 1-11.

Francis R A. (2011) Wall ecology: A frontier of urban biodiversity and ecological engineering. *Progress in Physical Geography*, 35, 43-63.

Fu W, Liu S, Degloria S D, et al. (2010) Characterizing the "fragmentation-barrier" effect of road networks on landscape connectivity: a case study in Xishuangbanna, Southwest China. *Landscape and Urban Planning*, 95, 122-129.

Fuller R A, Irvine K N, Devine-Wright P, et al. (2007) Psychological benefits of greenspace increase with biodiversity. *Biology Letters*, 3, 390-394.

Gago E J, Roldan J, Pacheco-Torres R, et al. (2013) The city and urban heat islands: A review of strategies to mitigate adverse effects. *Renewable and Sustainable Energy Reviews*, 25, 749-758.

Garland T. (1984) Physiological correlates of locomotory performance in a lizard: an allometric approach. *American Journal of Physiology*, 247, 806-815.

Garrard G, Bekessy S. (2014) Land use and land management. In *Australian Environmental Planning: Challenges and Future Prospects*. Byrne J, Sipe N, Dodson J eds. Routledge, Abingdon, 61-72.

Gaston K J, Fuller R A, Loram A, et al. (2007) Urban domestic gardens (XI): variation in urban wildlife gardening in the United Kingdom. *Biodiversity and Conservation*, 16, 3227-3238.

Gaston K J, Warren P H, Thompson K, et al. (2005) Urban domestic gardens (IV): the extent of the resource and its associated features. *Biological Conservation*, 14, 3327-3349.

Germano J M, Bishop P J. (2009) Suitability of amphibians and reptiles for translocation. *Conservation Biology*, 23, 7-15.

Gibbs J P, Shriver W G. (2002) Estimating the effects of road mortality on turtle populations. *Conservation Biology*, 16, 1647-1652.

Gibbs J P, Shriver W G. (2005) Can road mortality limit populations of pool-breeding amphibians? *Wetland Ecology and Management*, 13, 281-289.

Goddard M A, Dougill A J, Benton T G. (2010) Scaling up from gardens: biodiversity conservation in urban environments. *Trends in Ecology and Evolution*, 25, 90-98.

Goddard M A, Dougill A J, Benton T G. (2013) Why garden for wildlife? Social and ecological drivers, motivations and barriers for biodiversity management in residential land-scape. *Ecological Economics*, 86, 258-273.

Gonzalez-Garcia A, Belliure J, Gomez-Sal A, et al. (2009) The role of urban greenspaces in fauna conservation: the case of the iguana *Ctenosaura similis* in the "patios" of León city, Nicaragua. *Biodiversity and Conservation*, 18, 1909-1920.

Gordon A, Simondson D, White M, et al. (2009) Integrating conservation planning and landuse planning in urban landscapes. *Landscape and Urban Planning*, 91, 183-194.

Graham N A J, Cinner J E, Norstrom A V, et al. (2014) Coral reefs as novel ecosystems: embracing new futures. *Current Opinion in Environmental Sustainability*, 7, 9-14.

Grant B W, Middendorf G, Colgan M J, et al. (2011) Ecology of urban amphibians and reptiles: urbanophiles, urbanophobes, and the urbanoblivious. In *Urban Ecology: Patterns, Processes, and Applications*. Niemelä J ed. Oxford University Press, Oxford, 167-178.

Grewal S S, Grewal P S. (2012) Can cities become self-reliant in food? *Cities*, 29, 1-11.

Gupta R B, Chaudhari P R, Wate, S. R. (2008) Floristic diversity in urban forest area of. NEERI Campus, Nagpur, Maharashtra (India). *Journal of Environmental Science & Engineering*, 50, 55-62.

Hahs A K, McDonnell M J, McCarthy M A, et al. (2009) A global synthesis of plant extinction rates in urban areas. *Ecology Letters*, 12, 1165-1173.

Hale J M, Heard G W, Smith K L, et al. (2013) Structure and fragmentation of growling grass frog metapopulations. *Conservation Genetics*, 14, 313-322.

Hall T. (2011) What has happened to the great Aussie backyard? *The Conversation*. http: // theconversation. com/what-has-happened-to-the-great-aussie-backyard-

4506 (accessed 06/06/2015).

Hamer A J, van der Ree R, Mahony M J, et al. (2014) Usage rates of an under-road tunnel by three Australian frog species: implications for road mitigation. *Animal Conservation*, 17, 379-387.

Harlan S L, Brazel A J, Jenerette G D, et al. (2007) In the shade of affluence: the inequitable distribution of the urban heat island. *Research in Social Problems and Public Policy*, 15, 173-202.

Heard G W, McCarthy M A, Scroggie M P, et al. (2013) A Bayesian model of metapopulation viability, with application to an endangered amphibian. *Diversity and Distributions*, 19, 555-566.

Heard G W, Scroggie M P, Malone B S. (2012) The life history and decline of the threatened Australian frog, *Litoria raniformis*. *Austral Ecology*, 37, 276-284.

Hes D, Plessis C D. (2015) *Designing for Hope: Pathways to Regenerative Sustainability*. Routledge, Abingdon.

Hetrick S, Chowdhury R R, Brondizio E, et al. (2013) Spatiotemporal patterns and socioeconomic contexts of vegetative cover in Altamira City, Brazil. *Land*, 2, 774-796.

Hilty J A, Lidlicker W Z, Merenlender A M. (2006) *Corridor Ecology: The Science and Practice of Linking Landscapes for Biodiversity Conservation*. Island Press, Washington DC.

Hobbs R J, Higgs E, Hall C. (2013) *Novel Ecosystems: Intervening in the New Ecological World Order*. Wiley-Blackwell, Oxford.

Hobbs R J, Higgs E, Harris J A. (2009) Novel ecosystems: implications for conservation and restoration. *Trends in Ecological and Evolution*, 24, 599-605.

Hobbs R J, Arico S, Aronson J, et al. (2006) Novel ecosystems: theoretical and management aspects of the new ecological world order. *Global Ecology and Biogeography*, 15, 1-7.

Hunter A M, Williams N S G, Rayner J P, et al. (2014) Quantifying the thermal performance of green facades: A critical review. *Ecological Engineering*, 63,

102-113.

Hunter M R, Hunter M D. (2008) Designing for conservation of insects in the built environment. *Insect Conservation and Diversity*, 1, 189-196.

IAIA(International Association for Impact Assessment). (2002) Strategic environmental assessment performance criteria. *IAIA Special Publication Series*, 1. http://www.iaia.org/publicdocuments/special-publications/sp1.pdf? AspxAutoDetectCookieSupport=1(accessed 25/06/2015).

Indykiewicz P, Jerzak L, Bohner J, et al. (2011) Urban Fauna, in *Studies of Animal Biology, Ecology and Conservation in European Cities*. University of Technology and Life Sciences in Byd-goszcz, Bydgoszcz.

Ives C D, Lentini P E, Threlfall C G, et al. (2016) Cities are hotspots for threatened species. *Global Ecology and Biogeography*, 25, 117-126.

Jacobs B, Mikhailovich N, Delaney C. (2014) *Benchmarking Australia's Urban Tree Canopy: An i-Tree Assessment*. Institute for Sustainable Futures, University of Technology Sydney. http://202020vision.com.au/media/7141/benchmarking_australias_urban_tree_canopy.pdf(accessed 25/06/2015).

Jaganmohan M, Vailshery L S, Gopal D, et al. (2012) Plant diversity and distribution in urban domestic gardens and apartments in Bangalore, India. *Urban Ecosystems*, 15, 911-925.

James C. (2009) Australian homes are the biggest in the world. *ComSec Economic Insights*, November 30, 2009. https://img.yumpu.com/37485169/1/358x507/australian-homes-are-biggest-in-the-world-comsec.jpg(accessed 12/11/2015).

Jim C Y. (2014) Ecology and conservation of strangler figs in urban wall habitats. *Urban Ecosystems*, 17, 405-426.

Jim C Y, Chen S S. (2003) Comprehensive greenspace planning based on landscape ecology principles in compact Nanjing city, China. *Landscape and Urban Planning*, 65, 95-116.

Jo H. (2002) Impacts of urban greenspace on offsetting carbon emissions for middle Korea. *Journal of Environmental Management*, 64, 115-126.

Kadas G. (2002) Rare invertebrates colonizing green roofs in London. *Urban Habitats*, 4, 66-86.

Karvonen A. (2011) *Politics of Urban Runoff: Nature, Technology, and the Sustainable City*. MIT Press, Cambridge.

Kendal D, Williams N S G, Williams K J H. (2010) Harnessing diversity in gardens through individual decision makers. *Trends in Ecology and Evolution*, 25, 201-202.

Kendal D, Williams N S G, Williams K J H. (2012) Drivers of diversity and tree cover in gardens, parks, and streetscapes in an Australian city. *Urban Forestry and Urban Greening*, 11, 257-265.

Koehler S L, Gilmore D C. (2015) Translocation of the threatened Growling Grass Frog *Litoria raniformis*: a case study. *The Australian Zoologist*, 37, 321-336.

Köhler M. (2008) Green facades—a view back and some visions. *Urban Ecosystems*, 11, 423-436.

Kowarik I. (2011) Novel urban ecosystems, biodiversity, and conservation. *Environmental Pollution*, 159, 1974-1983.

Knapp S, Kuhn I, Schweiger O, et al. (2008) Challenging urban species diversity: contrasting phylogenetic patterns across plant functional groups in Germany. *Ecology Letters*, 11, 1054-1064.

Le Roux D S, Ikin K, Lindenmayer D B, et al. (2014) The future of large old trees in urban landscapes. *PLoS ONE* 9, e99403.

Lee K E, Williams K J H, Sargent L D, et al. (2015) 40-second green roof views sustain attention: the role of micro-breaks in attention restoration. *Journal of Environmental Psychology*, 42, 182-189.

Lehvävirta S. (2007) Non-anthropogenic dynamic factors and regeneration of (hemi) boreal urban woodlands-synthesising urban and rural ecological knowledge. *Urban Forestry and Urban Greening*, 6, 119-134.

Lenda M, Skorka P, Moron D, et al. (2012) The importance of the gravel excavation industry for the conservation of grassland butterflies. *Biological Conservation*, 148, 180-190.

Levick L R, Goodrich D C, Hernandez M, et al. (2008) *The Ecological and Hydrological Significance of Ephemeral and Intermittent Streams in the Arid and Semi-Arid American Southwest*. US Environmental Protection Agency, Office of Research and Development. http://www.epa.gov/esd/land-sci/pdf/EPHEMERAL_STREAMS_REPORT_Final_508-Kepner.pdf(accessed 25/06/2015).

Li F, Wang R, Paulussen J, et al. (2005) Comprehensive concept planning of urban greening based on ecological principles: a case study in Beijing, China. *Landscape and Urban Planning*, 72, 325-336.

Lindenmayer D B, Laurance W F, Franklin J F. (2012) Global decline in large old trees. *Science*, 338, 1305-1306.

Livesley S J, Dougherty B J, Smith A J, et al. (2010) Soil-atmosphere exchange of carbon dioxide, methane, and nitrous oxide in urban garden systems: impact of irrigation, fertiliser, and mulch. *Urban Ecosystems*, 13, 273-293.

Loram A, Warren P H, Gaston K J. (2008) Urban domestic gardens (XIV): the characteristics of gardens in five cities. *Environmental Management*, 42, 361-376.

Mahony M. (1999) Review of the declines and disappearances within the bell frog species group (*Litoria aurea* species group) in Australia. In *Declines and Disappearances of Australian Frogs*. Campbell A ed. Environment Australia, Canberra, 81-93.

Marchand M N, Litvaitis J A. (2004) Effects of habitat features and landscape composition on the population structure of a common aquatic turtle in aregion undergoing rapid development. *Conservation Biology*, 18, 758-767.

Maric T, Zaninovic J, Scitaroci B B O. (2008) Landscape as a connection-beyond boundaries. Schrenk M, Popovich V V, Zeile P, et al. eds. *REAL CORP* 2013: *Planning Times*. Proceedings of the 18th International Conference on Urban Planning, Regional Development and Information Society. CORP-Competence Center of Urban and Regional Planning, Schwechat, Austria, 497-506. http://corp.at/fileadmin/proceedings/CORP2013_proceedings.pdf(accessed 25/06/2015).

Marriage G. (2010) Minimum vs maximum: size and the New Zealand house. 2010 *Aus-*

tralasian Housing Researchers' Conference. http://www.academia.edu/3712795/Minimum_vs_Maximum_size_and_the_New_Zealand_House_-_first_published_in_the_Australasian_ Housing_Researchers_Conference_Auckland_2010(accessed 25/06/2015).

Martell M S, Englund J V, Tordoff H B. (2002) An urban osprey population established by translocation. *Journal of Raptor Research*, 36, 91-96.

McDonald R I, Kareiva P, Forman R T T. (2008) The implication of current and future urbanization for global protected areas and biodiversity conservation. *Biological Conservation*, 141, 1695-1703.

McKinney M L. (2002) Urbanization, biodiversity, and conservation. *BioScience*, 52, 883-890.

McPherson E G, Rowntree R A. (1993) Energy conservation potential of urban tree planting. *Journal of Arboriculture*, 19, 321-331.

McPherson E G, Nowak D, Heisler G, et al. (1997) Quantifying urban forest structure, function, and value: the Chicago Urban Forest Climate Project. *Urban Ecosystems*, 1, 49-61.

MEA(Millennium Ecosystem Assessment). (2005) *Ecosystems and Human Well-Being*, World Resources Institute, Washington DC.

Mentens J, Raes D, Hermy M. (2006) Green roofs as a tool for solving the rainwater runoff problem in the urbanized 21st century? *Landscape and Urban Planning*, 77, 217-226.

Milton S J. (2003) Emerging ecosystems: a washing-stone for ecologists, economists and sociologists? *South African Journal of Science*, 99, 404-406.

Mitchell J C, Jung Brown R E, Bartholomew B. (2008) *Urban Herpetology*. Herpetological Conservation 3. Society for the Study of Amphibians and Reptiles, Salt Lake City.

Murcia C, Aronson J, Kattan G H, et al. (2014) A critique of the "novel ecosystem" concept. *Trends in Ecology and Evolution*, 29, 548-553.

Newbound M, Bennett L T, Tibbits J, et al. (2012) Soil chemical properties, rather than landscape context, influence woodland fungal communities along an urban-rural gradient. *Austral Ecology*, 37, 236-247.

Ng C S L, Chen D, Chou L M. (2012) Hard coral assemblages on seawalls in Singapore. *Contributions to Marine Science*, 2012, 75-79.

Ng S J, Dole J W, Sauvajot R M, et al. (2004) Use of highway undercrossings by wildlife in southern California. *Biological Conservation*, 115, 499-507.

Norton B A, Bosomworth K, Coutts A, et al. (2013) *Planning for a Cooler Future: Green Infrastructure to Reduce Urban Heat.* Victorian Centre for Climate Change Adaptation Research, Melbourne. http://www.vcccar.org.au/sites/default/files/publications/VCCCAR%20Green%20Infrastructure%20Guide%20Final.pdf (accessed 25/06/2015).

Norton B A, Coutts A M, Livesley S J, et al. (2015) Planning for cooler cities: A framework to prioritise green infrastructure to mitigate high temperatures in urban landscapes. *Landscape and Urban Planning*, 134, 127-138.

NSW Department of Planning. (2010) *Sydney Growth Centres Strategic Assessment Program Report.* Department of Planning New South Wales, Sydney. http://www.environment.gov.au/epbc/notices/assessments/pubs/sydney-growth-centres-program-report.pdf(accessed 25/06/2015).

Oberndorfer E, Lundholm J, Bass B, et al. (2007) Green roofs as urban ecosystems: ecological structures, functions, and services. *BioScience*, 57, 823-833.

Ong B L. (2003) Green plot ratio: an ecological measure for architecture and urban planning. *Landscape and Urban Planning*, 63, 197-211.

Parris K M. (2006) Urban amphibian assemblages as metacommunities. *Journal of Animal Ecology*, 75, 757-764.

Parris K M. (2009) What are strategic impact assessments? *Decision Point* 32, 4-6. http://decision-point.com.au/wp-content/uploads/2014/12/DPoint_32.pdf (accessed 25/06/2015).

Pauchard A, Aguayo M, Peña E, et al. (2006) Multiple effects of urbanization on the biodiversity of developing countries: The case of a fast-growing metropolitan area

(Concepción, Chile). *Biological Conservation*, 127, 272-281.

Pedlowski M A, Da Silva V A C, Adell J J C, et al. (2002) Urban forest and environmental inequality in Campos dos Goytacazes, Rio de Janeiro, Brazil. *Urban Ecosystems*, 6, 9-20.

Pietsch R S. (1994) The fate of urban Common Brushtail Possums translocated to sclerophyll forest. In *Reintroduction Biology of Australian and New Zealand Fauna*. Serena M ed. Surrey Beatty and Sons, Chipping Norton, 239-246.

Polasky S, Nelson E, Camm J, et al. (2008) Where to put things? Spatial land management to sustain biodiversity and economic returns. *Biological Conservation*, 141, 1505-1524.

Polk M H, Young K R, Crews-Meyer K A. (2005) Biodiversity conservation implications of landscape change in an urbanizing desert of Southwestern Peru. *Urban Ecosystems*, 8, 313-334.

Powell P, Henderson R W. (2008) Urban herpetology in the West Indies. Mitchell J C, Jung Brown R E, Bartholomew B eds. *Urban Herpetology*. Herpetological Conservation 3. Society for the Study of Amphibians and Reptiles, Salt Lake City, 389-404.

Pyke G H. (2002) A review of the biology of the southern bell frog *Litoria raniformis* (Anura: Hylidae). *Australian Zoologist*, 32, 32-48.

Rilov G, Benayahu Y. (1998) Vertical artificial structures as an alternative habitat for coral reef fishes in disturbed environments. *Marine Environmental Research*, 45, 431-451.

Rosenzweig M L. (2003a) *Win-Win Ecology: How the Earth's Species Can Survive in the Midst of Human Enterprise*. Oxford University Press, Oxford.

Rosenzweig M L. (2003b) Reconciliation ecology and the future of species diversity. *Oryx*, 37, 194-205.

Rudd H, Vala J, Schaefer V. (2002) Importance of backyard habitat in a comprehensive biodiversity conservation strategy: a connectivity analysis of urban green spaces. *Restoration Ecology*, 10, 368-375.

Rumming K. (2004) *Hannover Kronsberg Handbook, Planning and Realisation*. City of Han-nover, Germany. Saggin, G. (2013) Urban beehives on the increase as global bee numbers decline. http://www.abc.net.au/news/2013-11-15/urban-beehive-movement-in-australia-and-around-the-world/5093764(accessed 25/06/2015).

Savard J L, Clergeau P, Mennechez G. (2000) Biodiversity concepts and urban ecosystems. *Landscape and Urban Planning*, 48, 131-142.

Seastedt T R, Hobbs R J, Suding K N. (2008) Management of novel ecosystems: are novel approaches required? *Frontiers in Ecology and the Environment*, 6, 547-553.

Secretariat of the Convention on Biological Diversity. (2012) *Cities and Biodiversity Outlook*. Secretariat of the Convention on Biological Diversity, Montreal. http://www.cbd.int/doc/health/cbo-action-policy-en.pdf(accessed 25/06/2015).

Seto K C, Guneralp B, Hutyra L R. (2012) Global forecasts of urban expansion to 2030 and direct impacts on biodiversity and carbon pools. *Proceedings of the National Academy of Sciences of the United States of America*, 109, 16083-16088.

Shanahan D F, Lin B B, Gaston K J, et al. (2014) Socio-economic inequality in access to nature on public and private lands: A case study from Brisbane, Australia. *Landscape and Urban Planning*, 130, 14-23.

Shariful Islam K M. (2004) Rooftop gardening as a strategy of urban agriculture for food security: the case of Dhaka City, Bangladesh. *Acta Horticulturae*, 643, 241-247.

Snyder S A, Miller J R, Skibbe A M, et al. (2007) Habitat acquisition strategies for grassland birds in an urbanizing landscape. *Environmental Management*, 40, 981-992.

Stagoll K, Lindenmayer D B, Knight E, et al. (2012) Large trees are keystone structures in urban parks. *Conservation Letters*, 5, 115-122.

Standish R J, Hobbs R J, Miller J R. (2013) Improving city life: options for ecological restoration in urban landscapes and how these might influence interactions between people and nature. *Landscape Ecology*, 28, 1213-1221.

Stovin V R, Jorgensen A, Clayden A. (2015) Street trees and stormwater management. *Arboriculture Journal: The International Journal of Urban Forestry*, 30, 297-310.

Sung C Y. (2013) Mitigating surface urban heat island by a tree protection policy: A case study of The Woodland, Texas, USA. *Urban Forestry and Urban Greening*, 12, 474-480.

Tanner C J, Adler F R, Grimm, N. B, et al. (2014) Urban ecology: advancing science and society. *Frontiers in Ecology and the Environment*, 12, 574-581.

Taylor L, Hochuli D F. (2015) Creating better cities: how biodiversity and ecosystem functioning enhance urban residents' wellbeing. *Urban Ecosystems*, 18, 747-762.

Threlfall C G, Law B, Banks P B. (2012) Influence of landscape structure and human modifications on insect biomass and bat foraging activity in an urban landscape. *PLoS ONE* 7, E38800.

Threlfall C G, Williams N S G, Hahs A K, et al. (2014) Green havens. *Australian Turfgrass Management* 16.5, September-October 2014. http://www.agcsa.com.au/files/Biodiversity%20165%20pg%206-12pdf.pdf (accessed 25/06/2015).

Tonietto R, Fant J, Ascher J, et al. (2011) A comparison of bee communities of Chicago green roofs, parks and prairies. *Landscape and Urban Planning*, 103, 102-108.

Tyrvainen L, Pauleit S, Seeland K, et al. (2005) Benefits and uses of urban forests and trees. In *Urban forests and Trees: A Reference Book*. Konijnendijk C C, Nilsson K, Randrup T B, et al. eds. Springer, Berlin, 81-114.

Tzoulas K, Korpela K, Venn S, et al. (2007) Promoting ecosystem and human health in urban areas using green infrastructure: a literature review. *Landscape and Urban Planning*, 81, 167-178.

Umwelt. (2013) *Gungahlin Strategic Assessment: Final Assessment Report*. Prepared on behalf of the ACT Economic Development Directorate and ACT Environment and Sustainable Development Directorate, Canberra. http://www.economicdevelopment.act.gov.au/__data/assets/pdf_file/0009/480177/8024_R02_V7-Assessment-Report.pdf (accessed 25/06/2015).

UN-HABITAT. (2010) *The State of African cities: Governance, Inequality and Urban Land Markets*. United Nations Human Settlement Program, Nairobi, Kenya.

http：//mirror. unhabitat. org/ pmss/listItemDetails. aspx? publicationID = 3034 （accessed 25/06/2015）.

Urbonas B R, Doerfer J T. (2005) Stream protection in urban watersheds through master planning. *Journal of Water Science and Technology*, 51, 239-247.

US Census Bureau. (2014) *Median and Average Square Feet of Floor Area in New Single-Family Houses Completed by Location*. US Census Bureau, Washington DC. https：//www. census. gov/construction/chars/pdf/medavgsqft. pdf （accessed 10/07/15）.

Uy P D, Nakagoshi N. (2008) Application of land suitability analysis and landscape ecology to urban greenspace planning in Hanoi, Vietnam. *Urban Forestry and Urban Greening*, 7, 25-40.

Vallance S, Perkins H C, Bowring J, et al. (2012) Almost invisible: glimpsing the city and its residents in the urban sustainability discourse. *Urban Studies*, 49, 1695-1710.

van der Grift E A, van der Ree R, Fahrig L, et al. (2013) Evaluating the effectiveness of road mitigation measures. *Biodiversity and Conservation*, 22, 425-448.

Vergnes A, Kerbiriou C, Clergeau P. (2013) Ecological corridors also operate in an urban matrix: A test case with garden shrews. *Urban Ecosystems*, 16, 511-525.

Vergnes A, Le Viol I, Clergeau P. (2012) Green corridors in urban landscape affect the arthropod communities of domestic gardens. *Biological Conservation*, 154, 171-178.

von Haaren C, Reich M. (2006) The German way to greenways and habitat networks. *Landscape and Urban Planning*, 76, 7-22.

Walsh C J, Roy A H, Feminella J W, et al. (2005) The urban stream syndrome: current knowledge and the search for a cure. *Journal of North American Benthological Society*, 24, 706-723.

Wang Y, Piao Z J, Wang X Y, et al. (2013) Road mortalities of vertebrate species on Ring Changbai Mountain Scenic Highway, Jilin Province, China. *North-Western Journal of Zoology*, 9, 399-409.

Ward A I, Dendy J, Cowan D P. (2015) Mitigation impacts of roads on wildlife: an agenda for the conservation of priority European protected species in Great Britain. *European Journal of Wildlife Research*, 61, 199-211.

Weeks J R, Hill A, Stow D, et al. (2007) Can we spot a neighbourhood from the air? Defining neighbourhood structure in Accra, Ghana. *Geojournal*, 69, 9-22.

White A W, Pyke G H. (2008) Frogs on the hop: translocations of Green and Golden Bell Frogs *Litoria aurea* in Greater Sydney. *Australian Zoologist*, 34, 249-260.

Williams N S G, Morgan J W, McCarthy M A, et al. (2006) Local extinction of grassland plants: the landscape matrix is more important than patch attributes. *Ecology*, 87, 3000-3006.

Williams N S G, Morgan J W, McDonnell M J, et al. (2005) Plant traits and local extinctions in natural grasslands along an urban-rural gradient. *Journal of Ecology*, 93, 1203-1213.

Wilson J N, Bekessy S, Parris K M, et al. (2013) Impacts of climate change and urban development on the spotted marsh frog (*Limnodynastes tasmaniensis*). *Austral Ecology*, 38, 11-22.

Windmiller B, Calhoun A J K. (2008) Conserving vernal pool wildlife in urbanizing landscapes. In *Science and Conservation of Vernal Pools in Northeastern North America*. Calhoun A J K, DeMaynadier P G eds. CRC Press, Boca Raton, 235-251.

Wolch J R, Byrne J, Newell J P. (2014) Urban green space, public health, and environmental justice: The challenge of making cities "just green enough". *Landscape and Urban Planning*, 125, 234-244.

Wong N H, Tay S F, Wong R, et al. (2003) Life cycle cost analysis of rooftop gardens in Singapore. *Building and Environment*, 38, 499-509.

Yasuda M, Koike F. (2006) Do golf courses provide a refuge for flora and fauna in Japanese urban landscapes? *Landscape and Urban Planning*, 75, 58-68.

Yencken D, Wilkinson D. (2000) *Resetting the Compass: Australia's Journey Towards Sustainability*. CSIRO Publishing, Melbourne.

第 8 章 总结和展望

8.1 概述

本书是城市生态学的导论。在前面的章节中,笔者应用发达国家和发展中国家的例子综合叙述城市环境的生态学,也理顺并研究了城镇化过程中发生的生物物理过程,并探讨了这些过程如何共同影响城市环境的特征(第2章),以及影响种群、群落和生态系统的动态(第3至5章)。笔者分析了城市人口生态学以及城市环境可以通过多种方式影响其居民的健康和福祉(第6章)。笔者已经论证了在城市环境中保护生物多样性和维护生态系统服务的目的,是为了人类和城市中其他物种的利益,并概述了实现这些目标的一些实用策略(第7章)。现在剩下的是回答一个首要问题,并讨论城市生态学的未来发展方向。

8.2 我们是否需要一种新的城市生态学理论?

首要问题是,我们是否需要一种新的城市生态学理论,还是可以使用现有的生态学理论来理解城市环境生态学?城市生态学的一支学派认为,当前的生态学理论无法捕捉到自然系统和人类系统相交的城市生态系统的全部复杂性,有时被称为人类生态系统或社会生态系统耦合(Liu et al., 2007; Alberti, 2008; Collins et al., 2011)。例如,Alberti(2008)提出:"城市生态系统展现出独特的属性、模式和行为,这些都是由人类与生态过程之间的复杂耦合产生的",现有生态理论不足以理解其功能和动态。此外,她坚持认为,我们需要一种城市生态学理论,该理论以多个学科的多个概念为基础,用以解决城市生态系统行为的潜在机制(Alberti,2008)。Pickett 等(2008)指出,城市生态学缺乏理论,而 Cadenasso 和 Pickett(2008)则稍加调整,表明城市生态学还没有"完整、成熟"的理论。除了完整和不完整理论之间的区别之外,这些作者针对城市生态系统的复杂性、人类主导和独特性提出许多有趣的观点,将在下面进行探讨。

8.2.1 城市生态系统的复杂性

有人认为,城市生态系统比其他类型的生态系统更复杂,这是因为它们具有动态性、异质性(跨多个时空尺度)、非线性、人口活动的多样性,以及城市环境中存在的社会生物物理反馈(Alberti et al.,2003;2008;Andersson,2006;Pickett et al.,2008;Qureshi et al.,2014)。例如,Alberti 等(2003)提出,城市研究人员需要明确"解决许多因素的复杂性,这些因素在从个体到区域、全球范围同时进行"。但是,这些特征均不局限于城市生态系统。许多其他生态系统也是动态的和异质性的,具有不遵循线性轨迹的各种特性和过渡状态(Levin,1998;Pen-Mouratov,Steinberger,2005;Anand et al.,2010)。在多个时空尺度上同时运行的许多变量的复杂性,在生态学领域并没有什么不同(Levin,1998;Anand et al.,2010);这种复杂性可以从沙漠、草原、林地、溪流和湖泊,到礁石和大洋的各种生态系统中观察到(George et al.,1992;Pen-Mouratov,Steinberger,2005;Hobbs et al.,2007;Boit et al.,2012;Sugihara et al.,2012;Bozec et al.,2013;Laurance et al.,2014)。

现在,地球上的每个生态系统都受到某种程度的人类活动的影响(Vitousek et al.,1997),或直接(如人类采集植物获取食物和药草、狩猎野生动物,或通过农业和生产林业改变景观特征),或间接(如人为气候变化及其对极地冰盖和北方冻土带永久冻土的影响)(Wipf et al.,2009;Schuur et al.,2015)。人类在所有生态系统中的行为都受到多种复杂而动态的社会因素影响,并且每个生态系统都以随时间和空间变化的方式对人类的行为作出响应。鉴于我们非常了解人类动机和活动,无论是个人还是社会集体,甚至可以说城市生态系统的复杂性要比其他物种占主导地位的生态系统的复杂性低。笔者认为过分强调城市生态系统的复杂性,可能会阻碍而不是帮助我们加深对城市、环境和居民的了解。

8.2.2 人类对城市生态系统的控制

如上所述,人类对生态系统的影响是连续的,而不是二元变量,只是程度问题,而不是"非此即彼"的命题。城市生态系统显然处于人类影响梯度的右端,但是人类的主导地位及其在城市中的各种行为是否将城市与所有其他生态系统区分开来?正如Collins等(2000)所述,城市中人类的活动以多种重要方式影响着城市生态系统,包括改变景观、调动养分、转移水分和能源、改变当地气候、驱动某些物种灭绝,同时又增加了其他物种的数量。然而,如果人类确实是大自然及其居住生态系统的组成部分(McDonnell, Pickett, 1997),那么他们在塑造这些生态系统方面的行动,不应该与塑造其他生态系统的非人类物种的行为有所不同,非人类物种在塑造如沙漠、湿地或珊瑚礁等其他类型的生态系统方面发挥着重要作用。相反,人类可以被视为城市中的生态系统工程师(Jones et al., 1994; Adler, Tanner, 2013),通过改变环境以适应自身的方式来影响景观结构,进而影响其他物种(Hastings et al., 2007)。

生态系统工程师是通过引起生物或非生物材料的物理状态变化,而直接或间接调节其他物种资源可用性的生物(Jones et al., 1994)。生态系统工程师的著名例子包括在河流上筑坝、土拨鼠挖洞以及聚集形成大片海底珊瑚礁的珊瑚虫(Wright et al., 2002; van Nimwegen et al., 2008; Wild et al., 2011; Bozec et al., 2013)。可以将人类归为同种异体的生态系统工程师,他们可以通过将物质从一种物理状态转换为另一种物理状态来改变环境(Jones et al., 1994)。相反,诸如珊瑚之类的自生工程师则通过自生的物理结构(躯体)来改变环境。在非城市生态系统中存在一个占主导地位的生态系统工程师,并不意味着无法应用现有的生态理论来解释该生态系统的生态。相反,工程师是该理论的一部分(Jones et al., 1997)。同样,没有理由说明人类作为城市占主导地位的生态系统工程师,就应该使城市环境的生态发生如此大的变化,以至于现有的生态理论不再适用城市环境。

8.2.3 城市生态系统的独特性

城市生态系统真的独特吗?在某些方面,论点成立——高覆盖率的不透水表面的，高密度的人口和建筑结构，高水平的大气、噪声和光的污染都使城市生态系统与其他生态系统区分开来。城市景观在空间上也往往是异质性的，但这本身不足以将它们与非城市景观区分开。如第7章所述，一些城市生态系统可能代表新型生态系统，并结合了新的环境条件和栖息地特征(如建筑物、道路廊道和人工防波堤)，从而支持由本土的和引进物种组成的新的生态群落(Hobbs et al.，2006；Standish et al.，2013)。然而，所有生态系统在某种程度上都是独特的。例如，森林因树木密度高与其他生态系统区分开来，热带雨林与温带落叶林因当地环境变量(降雨量、温度、湿度、土壤养分和水分含量)，植被的结构形式以及栖息于每种森林类型的植物、动物、真菌和细菌物种的组合而有所区别。

城市生态系统的独特性并不意味着它们具有独特的生态(Niemelä，1999；Catterall，2009)，城市生态系统的新颖性也并不意味着我们需要一种新颖的生态学理论来理解它们。城市环境中的所有类型的种群(包括人类种群)与其他类型环境中的种群一样，都经历着相同的出生、死亡、迁入和迁出过程(第3章)。城市生态群落同样是由选择、扩散、生态漂移和进化多样化过程形成的(第4章)，如同在沙漠、草原、森林、溪流和海洋中的过程。尽管由于城市建设和运营而造成破坏，但城市生态系统仍可以循环碳、水和养分，并提供诸如初级生产、吸收大气污染物、遮阴、降温、对害虫进行生物防治以及废物处理等服务(第5至7章)。为了应对Collins等(2000)的挑战，经典生态学理论可以解释城市中生物扩散、多度以及生物与环境之间的关系。许多人类因素的存在、他们的行为、偏好、动机、权力结构和社会经济条件，都会在各种时空尺度上影响城市景观的生物物理特征，以及城市中非人类物种的特性、分布和多度。但是，它们不会改变城市环境的基本生态。

8.3 城市生态学的定义和范围

如果正如笔者所说的那样，现有的生态学理论足以解释城市环境生态学组织的各个层面，从个体生物到种群、生态群落和整个城市生态系统，那么为什么还要继续呼吁建立新的城市生态学理论呢？答案可能在于那些反对现有生态学理论的适用性和充分性的人所使用的城市生态学定义。例如，Marzluff等(2008)将城市生态学描述为一个新兴的跨学科领域，它在生态学、社会学、地理学、城市规划、工程学、经济学、公共卫生和人类学等众多学科中都有深厚的渊源。McDonnell(2011、2015)提出，城市生态学正在成为一种真正的跨学科科学，它借鉴了许多不同的自然科学和社会科学；而 Wu(2014)则宣称"……城市生态学已经发展成为一个真正跨学科科学，它融合了生态学、地理学、规划学和社会学"。

但是，从分类的角度来看，一个实体不能由它自己加上一系列其他实体组成。因此，城市生态学在逻辑上不能包括城市生态学加上城市规划、社会学、地理学等，仍被归类为城市生态学。这种跨学科的集聚范围必须更大。笔者建议将其称为城市科学，类似于农业科学或森林科学。农业科学涵盖与农业生态系统的发展、结构、功能、动态和管理相关的多种学科，包括农业生态学、土壤学、农学、遗传学、畜牧业、昆虫学、植物病理学、经济学、文化地理学和社会学。同样，森林科学包括森林生态、火灾生态、土壤学、造林、遗传学、水文学、木材学、森林规划、采伐、社会学和经济学等多种学科。城市科学可以被视为一个相应的跨学科领域，它整合了来自各个学科(包括城市生态学)的学术成果，以研究城市环境的发展、动态、功能和管理；城市居民的动机、行动、健康和福祉；以及城镇化对生态系统和城市所有物种的影响。

8.4 我们需要一种新的城市科学理论吗？

现有的生态学理论不足以解释农业环境及其管理的各个方面(包括其物理、生物、经济、政治和社会驱动因素、过程、反馈和相互依赖)，现有的生态学理论也不足以解释森林环境管理的所有方面(提供木材、娱乐休闲、养护和其他生态系统服务)。因此，现有的生态理论也没能解释城市环境建设的各个方面，从城镇化的政治、社会和文化驱动因子，到城市系统的设计、规划和管理。这意味着我们需要一个新的城市科学理论，可以解释城镇化和城市系统的各个方面，即一切事物的城市理论？

笔者认为事实并非如此。在任何类型的生态系统中进行跨学科和学科内的研究，都可以借鉴与特定问题相关的每个组成学科中的现有理论。没有必要开发新的、通用的理论，来解答城市领域中所有可能学科的每个可能问题。例如，让我们考虑以下问题：在给定的城市中，我们如何设计、创建和管理公用绿地，以最大限度地提高它们在生物多样性、空气质量、微气候降温以及人类健康和福祉方面的综合利益？为了解决这个问题，我们需要广泛的学科投入，包括生态学、保护生物学、气候学、大气化学、植物生理学、气候学、大气化学、植物生理学、水文学、人口健康、健康提升和教育、环境心理学、社会学、城市规划、经济学和公园管理。每个学科都有自己的理论基础，可以用来构建整个问题的相关部分；当然，将现有的理论用于研究城市领域内的跨学科问题，比寻求重新发明理论更有效，甚至可能更强大。一些旨在结合多个学科的数据、方法和理论的概念框架已经存生（Collins et al., 2011）。尽管跨所有相关学科的综合理论可能是城市科学中一个有益的长期目标，但它并不是有效地规划和管理城市的生物多样性、改善生态系统功能或人类健康和福祉的必要条件。

8.5 未来方向

城市生态学未来的发展方向是什么？如果我们接受可以使用现有的生态学理论，来解决和改善城市环境的生态问题，那么这将使我们能够将精力集中在一系列有趣的城市生态问题上，例如：

1. 我们能否根据其生态学或生理特征来预测城镇化的各种生物物理过程对分类群中不同物种的影响？
2. 城镇化的生物物理过程如何影响人类和其他物种的种群过程（包括出生、死亡、迁入和迁出）？
3. 在城市环境中，物种的营养位置对其持久性或局部灭绝的概率有什么影响？
4. 这是否取决于城镇化前后生态群落的整体营养结构？
5. 扩散在维持城市多样化的生态群落中扮演什么角色？
6. 我们必须维持多少栖息地的连通性，以确保有机体在城市景观中的充分扩散？
7. 城市环境的某些特征是否正在推动非人类物种（从微生物、植物到动物）

的进化?

8. 人类作为生态系统工程师,其行为通过什么机制以及在多大程度上影响了非人类种群的持续性、生态群落组成以及城市生态系统的功能?

9. 这些机制在全球不同类型的城市和不同生物群落中,是否具有同等重要的作用?

10. 我们能否更好地理解这些机制,以扭转(在现有的城市景观中)或预防(在未来的城市景观中)城镇化对生物多样性、生态系统功能和生态系统服务的某些不良影响?

11. 城市中新的环境压力源,如交通噪声、人工夜景照明以及空气、水、土壤的污染,如何影响陆地、淡水和海洋栖息地中的个体、种群和生态群落?

12. 我们能在多大程度上扭转城镇化对生态群落的不利影响?

在城市科学更广泛领域内,还有许多有趣的问题需要解决,例如:

1. 我们如何更好地将城市生态学纳入城市规划和设计中,以增强城市环境中的生物多样性和生态系统服务?

2. 我们如何利用生物多样性和绿色基础设施来改善城市居民与大自然以及彼此之间联系,并克服由于生态系统服务分配不均而造成的环境不公?

3. 我们可以做哪些小规模的改变来改善现有城市社区的宜居性?

4. 我们如何建设未来的城市,为所有人类居民都能享有适当的生活水平,同时还要考虑其他物种的需求,并保持有价值的生态系统功能和服务?

使用我们已有的理论(如生态学以及其他相关理论,视情况而定)来解决诸如此类的问题,这比继续寻找万能的城市理论能更有效地推动城市生态学和城市科学的跨学科领域。鉴于事关重大——城市环境中生物多样性(包括许多受威胁物种)的持续性、城市生态群落结构、本地和全球范围内生态系统的功能以及数十亿生活在世界各地城市中的人的健康和福祉——获得有效的进步才是至关重要的。我们没有时间可以浪费。

思考题

1. 描述城市生态系统区别于其他类型的生态系统的特征。

2. 人类对城市环境的控制会改变其基本生态吗？

3. 你认为我们需要一种新的城市生态学理论吗？为什么或者为什么不？

4. 城市科学是描述城市跨学科（包括城市生态学、社会学、地理学、城市规划和管理学）研究的有用术语吗？可以使用哪些术语，它们的相对优点是什么？

5. 你认为在未来几十年中，城市科学领域的哪些学科视角的组合对改善人类和非人类种群的福祉最重要？

本章参考文献

Adler F R, Tanner C J. (2013) *Urban Ecosystems: Ecological Principles for the Built Environment*. Cambridge University Press, Cambridge.

Alberti M, Marzluff J M, Shulenberger E, et al. (2003) Integrating humans into ecology: opportunities and challenges for studying urban ecosystems. *BioScience*, 53, 1169-1179.

Alberti M. (2008) *Advances in Urban Ecology: Integrating Humans and Ecological Processes in Urban Ecosystems*. Springer, New York.

Anand M, Gonzales A, Guichard F, et al. (2010) Ecological systems as complex systems: challenges for an emerging science. *Diversity*, 2, 395-410.

Andersson E. (2006) Urban landscapes and sustainable cities. *Ecology and Society*, 11, 34.

Boit A, Martinez N D, Williams R J, et al. (2012) Mechanistic theory and modelling of complex food-web dynamics in Lake Constance. *Ecology Letters*, 15, 594-602.

Bozec Y, Yakob L, Bejarano S, et al. (2013) Reciprocal facilitation and non-linearity maintain habitat engineering on coral reefs. *Oikos*, 122, 428-440.

Cadenasso M L, Pickett S T A. (2008) Urban principles for ecology landscape design and management: scientific fundamentals. *Cities and the Environment*, 1, 1-16.

Cadenasso M L, Pickett S T A, Schwarz K. (2007) Spatial heterogeneity in urban ecosystems: reconceptualizing land cover and a framework for classification. *Frontiers in Ecology and the Environment*, 5, 80-88.

Catterall C P. (2009) Responses of faunal assemblages to urbanisation: Global research paradigms and an avian case study. In *Ecology of Cities and Towns*: *A Comparative Approach*. McDonnell M J, Hahs A K, Brueste J eds. Cambridge University Press, Cambridge, 129-155.

Collins J P, Kinzig A, Grimm N B, et al. (2000) A new urban ecology: modelling human communities as integral parts of ecosystems poses special problems for the development and testing of ecological theory. *American Scientist*, 88, 416-425.

Collins S L, Carpenter S R, Swinton S M, et al. (2011) An integrated conceptual framework for long-term social-ecological research. *Frontiers in Ecology and the Environment*, 9, 351-357.

George M R, Brown J R, Clawson W J. (1992) Application of nonequilibrium ecology to management of Mediterranean grasslands. *Journal of Range Management*, 45, 436-440.

Hastings A, Byers J E, Crooks J A, et al. (2007) Ecosystem engineering in space and time *Ecology Letters*, 10, 153-164.

Hobbs R J, Yates S, Mooney H A. (2007) Long-term data reveal complex dynamics in grassland in relation to climate and disturbance. *Ecological Monographs*, 77, 545-568.

Hobbs R J, Arico S, Aronson J, et al. (2006) Novel ecosystems: theoretical and management aspects of the new ecological world order. *Global Ecology and Biogeography*, 15, 1-7.

Jones C G, Lawton J H, Shachak M. (1994) Organisms as ecosystem engineers. *OIKOS*, 69, 373-386.

Jones C G, Lawton J H, Shachak M. (1997) Positive and negative effects of organisms as physical ecosystem engineers. *Ecology*, 78, 1946-1957.

Laurance W F, Andrade A S, Magrach A, et al. (2014) Apparent environmental synergism drives the dynamics of Amazonian forest fragments. *Ecology*, 95, 3018-3026.

Levin S A. (1998) Ecosystems and the biosphere as complex adaptive systems. *Ecosystems*, 1, 431-436.

Liu J, Dietz T, Carpenter S R, et al. (2007) Complexity of coupled human and natural systems Science, 317, 1513-1516.

Luo J, Wei Y H D. (2009) Modeling spatial variations of urban growth patterns in Chinese cities: the case of Nanjing. *Landscape and Urban Planning*, 91, 51-64.

Luck M, Wu J. (2002) A gradient analysis of urban landscape pattern: a case study from the Pheonix metropolitan region, Arizona, USA. *Landscape Ecology*, 17, 327-339.

Marzluff J M, Shulenberger E, Simon U, et al. (2008) An introduction to urban ecology as an interaction between humans and nature. In *Urban Ecology: An International Perspective on the Interaction Between Humans and Nature.* Marzluff J M, Shulenberger E, Endlicher W, et al. eds. Springer, New York, vii-xi.

McDonnell M J. (2011) The history of urban ecology. In *Urban Ecology: Patterns.* Niemelä J ed. *Processes and Applications.* Oxford University Press, Oxford, 5-13.

McDonnell M J. (2015) Journal of Urban Ecology: Linking and promoting research and practice in the evolving discipline of urban ecology. *Journal of Urban Ecology*, 1, 1-6.

McDonnell M J, Pickett S T A. (1997) *Humans as Components of Ecosystems: The Ecology of Subtle Human Effects and Populated Areas.* Springer-Verlag, New York.

Niemelä J. (1999) Is there a need for a theory of urban ecology? *Urban Ecosystems*, 3, 57-65.

Pen-Mouratov, S, Steinberger Y. (2005) Spatio-temporal dynamic heterogeneity of nematode abundance in a desert ecosystem. *Journal of Nematology*, 37, 26-36.

Pickett S T A, Cadenasso M L, Grove J M, et al. (2008) Beyond urban legends: an emerging framework of urban ecology, as illustrated by the Baltimore Ecosystem Study. *BioScience*, 58, 139-150.

Qureshi S, Haase D, Coles R. (2014) The Theorized Urban Gradient (TUG) method—a conceptual framework for socio-ecological sampling in complex urban agglomerations. *Ecological Indicators*, 36, 100-110.

Schuur E A G, McGuire A D, Schadel C, et al. (2015) Climate change and the permafrost carbon feedback. *Nature*, 520, 171-179.

Standish R J, Hobbs R J, Miller J R. (2013) Improving city life: options for ecological restoration in urban landscapes and how these might influence interactions between people and nature. *Landscape Ecology*, 28, 1213-1221.

Sugihara G, May R, Ye H, et al. (2012) Detecting causality in complex ecosystems. *Science*, 338, 496-500.

Van Nimwegen R E, Kretzer J, Cully J F Jr, (2008) Ecosystem engineering by a colonial mammal: how prairie dogs structure rodent communities. *Ecology*, 89, 3298-3305.

Vitousek P M, Mooney H A, Lubchenco J, et al. (1997) Human domination of Earth's ecosystems. *Science*, 277, 494-499.

Wild C, Hoegh-Guldberg O, Naumann M S, et al. (2011) Climate change impedes scleractinian corals as primary reef ecosystem engineers. *Marine and Freshwater Research*, 62, 205-215.

Wipf S, Stoeckli V, Bebi P. (2009) Winter climate change in alpine tundra: plant responses to changes in snow depth and snowmelt timing. *Climate Change*, 94, 105-121.

Wright J P, Jones C G, Flecker A S. (2002) An ecosystem engineer, the beaver, increases species richness at the landscape scale. *Oecologia*, 132, 96-101.

Wu J. (2014) Urban ecology and sustainability: the state-of-the-science and future directions. *Landscape and Urban Planning*, 125, 209-221.